PR**
SOME WE LOV
SOME WE EAT

"Hal Herzog does for our relationships with animals what Michael Pollan in *The Omnivore's Dilemma* did for our relationships with food. Presenting cutting-edge research and real-world stories with wit, sophistication, and charm, Herzog shows us that the relationships we have with animals, even the ones that just a minute ago seemed so sensible, are riddled with contradictions and complexities. The book is a joy to read, and no matter what your beliefs are now, it will change how you think."
—Sam Gosling, author of *Snoop: What Your Stuff Says About You*

"Herzog argues that moral absolutes are not readily available in a complex world—one that exists in shades of gray, rather than the black and white of animal rights activists and their opponents. . . . Herzog has a clear eye for the essence of a scientific study, but he leavens his narrative with illuminating personal stories and self-deprecating humor."
—*Nature*

"*Some We Love, Some We Hate, Some We Eat* is one of a kind. I don't know when I've read anything more comprehensive about our highly involved, highly contradictory relationships with animals—relationships that we mindlessly, placidly continue no matter how irrational they may be. Readers will welcome Herzog's eye-opening discussions, presented with compassion and humor. This page-turning book is quite something. You won't forget it anytime soon."
—Elizabeth Marshall Thomas, author of *The Hidden Life of Deer: Lessons from the Natural World*

"Everybody who is interested in the ethics of the relationships between humans and animals should read this book."

—Temple Grandin, author of *Animals Make Us Human*

"An instant classic. . . . Written so accessibly and personally, while simultaneously satisfying the scholar in all of us. . . . A smart and provocative book that is also a quick, enjoyable, and easy read, which is no easy task, given the complexity of the book's question and the diversity of the research drawn upon."

—Arnold Arluke, *Anthrozoös*

"*Some We Love, Some We Hate, Some We Eat* is both educational and enjoyable, a page-turner that I daresay puts Herzog in the same class as Malcolm Gladwell and Michael Lewis. Read this book. You'll learn some, you'll laugh some, you'll love some."

—John T. Slania, BookPage.com

"This is a wonderful book—wildly readable, funny, scientifically sound, and with surprising moments of deep, challenging thoughts. I loved it."

—Robert M. Sapolsky, neuroscientist at Stanford University and author of *Monkeyluv* and *A Primate's Memoir*

"Reminiscent of *Freakonomics*. . . . An agreeable guide to popular avenues of inquiry in the field of anthrozoology."

—*The New Yorker*

"Wonderful. . . . An engagingly written book that only seems to be about animals. Herzog's deepest questions are about men, women, and children."

—Karen Sandstrom, *Cleveland Plain Dealer*

"Hal Herzog deftly blends anecdote with scientific research to show how almost any moral or ethical position regarding our relationships with animals can lead to absurd consequences. In an utterly appealing narrative, he reveals the quirky—and for the reader, entertaining—ways we humans try to make sense of these absurdities."

—Irene M. Pepperberg, author of *Alex & Me: How a Scientist and a Parrot Discovered a Hidden World of Animal Intelligence—and Formed a Deep Bond in the Process*

"Fascinating. . . . Hal Herzog looks at the wild, tortured paradoxes in our relationship with the weaker, if sometimes more adorable, species."

—Kerry Lauerman, Salon.com

"Herzog writes about big ideas with a light touch. . . . Insightful, compassionate, and humorous."

—*Kirkus Reviews*

"Herzog delivers provocative popular science at its witty 'gee-whiz' best. With headings such as 'Feeding Kittens to Boa Constrictors,' the book challenges the reader to think through the knotty ethics of human interactions with other species. While it might make you squirm, you'll have fun reading this informal, often humorous survey of the emerging interdisciplinary field of anthrozoology."

—*Columbus Dispatch*

"A fun read. . . . What buoys this book is Herzog's voice. He's an assured, knowledgeable, and friendly guide."

—Associated Press

"Entertaining. . . . [Herzog] skillfully weaves research studies and personal anecdotes into a tightly knit overview of this exotic field."

—Phillip Manning, *Charlotte Observer*

"The thing that makes Hal Herzog's new book stand out from any number of other recent books on the subject is its dispassion. He's a professor in the emergent field of 'anthrozoology,' and he brings an anthropologist's unprejudiced attitude to his research. . . . The calm, reasoned tone of Herzog's book gives one room to think and invites readers to make up their own minds."

—*Independent Weekly* (North Carolina)

"An intelligent and amusing book that invites us to think deeply about how we define—and where we limit—our empathy for animals."

—*Publishers Weekly*

"Engaging and pleasantly cerebral. . . . When [Herzog is] talking to people about their views, the book is fascinating."

—*Time Out* (Chicago)

"Hal Herzog embraces the complexities of our relationships with nonhuman animals and helps us make sense of the contradictions. *Some We Love, Some We Hate, Some We Eat* will likely turn some die-hard meat eaters into vegetarians even as it convinces plenty of longtime vegans to give up their restrictive diets and enjoy a rare filet mignon. Herzog's dismissal of easy answers to difficult questions is a page-turning relief filled with an intriguing cast of characters few of us would otherwise meet."

—Peter Laufer, author of *The Dangerous World of Butterflies* and *Forbidden Creatures*

"Our relationships with nonhuman animal beings [are] confused, complicated, frustrating, and paradoxical. Hal Herzog captures the essence of our inability to think straight about other animals in a provocative book that should be required reading for anyone interested in trying to figure out who 'they' are and who we are. Read this book, read it again, and share it widely. It is *that* important."

—Marc Bekoff, author of The Emotional Lives of
Animals and The Animal Manifesto: Six Reasons for
Expanding Our Compassion Footprint and editor of the
Encyclopedia of Human-Animal Relationships

"Professor Hal Herzog writes lucidly. . . . No matter which side of the question you find yourself on, this book is illuminating, and dare I say quite entertaining."

—Sun Journal (North Carolina)

SOME WE LOVE,

SOME WE HATE,

SOME WE EAT

SOME WE LOVE, SOME WE HATE, SOME WE EAT

Why It's So Hard to Think Straight About Animals

SECOND EDITION

HAL HERZOG

HARPER PERENNIAL

NEW YORK • LONDON • TORONTO • SYDNEY • NEW DELHI • AUCKLAND

HARPER ● PERENNIAL

HarperCollins books may be purchased for educational, business, or sales promotional use. For information, please email the Special Markets Department at SPsales@harpercollins.com.

Designed by Jen Overstreet

Library of Congress Cataloging-in-Publication Data has been applied for.

ISBN 978-0-06-311928-4

23 24 25 26 27   CD   10 9 8 7 6 5 4 3 2

To the founders of the International Society for Anthrozoology: John Bradshaw, Erika Friedmann, Ben Hart, Lynette Hart, James Serpell, and Dennis Turner.

CONTENTS

I wrote *Some We Love, Some We Hate, Some We Eat* with two imaginary readers whispering in my ears. The first was my sister. She is an intellectually curious person who is not crazy about animals but is always on the lookout for a book that will hold her interest on a long plane flight. My second imaginary reader was one of my fellow researchers who has a skeptical streak. I wanted the book to be entertaining enough to keep the attention of a casual reader like my sister yet have sufficient scholarly chops that it would pass muster with my skeptical colleague. I was not sure I could pull it off. But I thought I'd give it a shot. After all, many books that have deeply influenced my thinking were written for a broad audience by academics—Richard Dawkins' *The Selfish Gene*, Steven Pinker's *The Blank Slate: The Modern Denial of Human Nature*, Peter Singer's *Animal Liberation: The Definitive Classic of the Animal Movement*, and Jonathan Haidt's *The Happiness Hypothesis: Finding Modern Truth in Ancient Wisdom*.

In some ways, my efforts were successful. When *Some We Love* appeared ten years ago, it garnered favorable reviews in scientific outlets such as *Nature* and also in *The New Yorker* and even *People* magazine. Readers on both sides of contentious animal ethics debates seemed to approve, though they did not always agree with me. The book has also had an impact on the study of human-animal interactions. While not primarily intended for researchers, it is regularly cited in journal articles, and it has been used in college courses ranging from freshman English classes to graduate school seminars.

Several years ago, students began asking me when I was going to write a new edition. My response was, "Never." I changed my mind when I used the book for a course I taught at the Osher Lifelong Learning Institute at the University of North Carolina in Asheville. I found my students were right; it was time for an update. To that end, I returned to my two imaginary readers. For readers like my sister, I have kept the backbone of the book. These are the stories of the animal people I have encountered over the years—the animal rights activists, the moral philosophers, the everyday pet-lovers, the biomedical researchers, the slaughterhouse workers, the dog show judges, and the turtle rescuers. For my fellow investigators, I bring in recent advances in research, including a couple hundred new citations in the reference section.

The scientific study of human-animal interactions goes back roughly forty years. But interest in the field among both animal people and researchers has skyrocketed over the last decade. Nearly fifty colleges and universities now offer degree programs in the field, and research centers have been established in North America, Europe, Asia, and Australia. A half dozen journals are devoted to human-animal studies, and thousands of research papers are published on the topic each year. As witnessed by extensive media coverage and a slew of trade books, the public seems to have an insatiable appetite for the latest findings on topics like the impact of pets on human health, the association between animal cruelty and human-directed violence, and the effectiveness of animal-assisted therapies.

While there are important updates in this edition of *Some We Love*, some aspects of the book remain the same. For the most part, I have not changed my mind about the major ideas, and I have tried to keep the conversational style of the first edition. Throughout the book, however, I have updated basic facts—the amount of money we spend on pets, shifts in public opinion on animal research, the percent of vegans who go back to eating animals. This edition also covers recent findings in research on the psychology of human-animal interactions. Whole areas of research have exploded in the past ten years. Between 2010 and 2020, for instance, the annual number of papers published on the use of therapy dogs tripled. Social psychologists are now pushing the boundaries by investigating topics such as how people negotiate the moral territory between loving animals and eating them, and the roles of emotion and logic in ethical decision making. The study of human-dog relationships has been a spectacularly fruitful area of investigation, and canine science research centers are cropping up worldwide.

The field has benefited from an influx of enthusiastic new investigators with sophisticated methodological skills. And, as in many areas of the social and biomedical sciences, more and better studies make for conflicting, sometimes controversial results. For example, new research indicates that living with pets does not necessarily improve the mental or physical health of humans, and the link between childhood animal cruelty and later adult violence turns out to be weaker than most people think. Other recent findings in this edition are related to the humanization of pets, the global uptick in the consumption of animals, shifts in strategies of the animal rights movement, and the ethics of animal research in response to the COVID-19 pandemic.

Some of the material in this edition of *Some We Love* originally appeared on the blog I write for *Psychology Today* magazine—*Animals and Us.* The blog now has over 150 posts which have been read over six million times. The posts are based on the latest research on human-animal interactions, and they deal with topics ranging from the role of genes in our love for pets to the ethics of de-sexing dogs.

When *Some We Love* was first published, I was interviewed for North Carolina Public Radio's *The State of Things*, hosted by Frank Stasio. When I walked into his office, Frank looked up and said, "Ah, Dr. Herzog. I just finished your book. It's not really about animals, is it." I wanted to hug the guy. He got it. At one level this book is about how we think and behave toward the creatures we share the world with—the ones we love, the ones we hate, the ones we eat. But the overriding theme is what our

INTRODUCTION

Why Is It So Hard to Think Straight About Animals

> I like pondering our relationships with animals because they tell a lot about who we are.
>
> —MARC BEKOFF, ETHOLOGIST

The way we think about other species often defies logic. Consider Judith Black. When she was twelve, Judith decided that it was wrong to kill animals just because they taste good. But what exactly is an animal? While it's obvious that dogs and cats and cows and pigs are animals, it was equally clear to Judith that fish were not. They just didn't *seem* like animals to her. For the next fifteen years, her intuitive biological classification system enabled Judith, who has a PhD in anthropology, to think of herself as a vegetarian yet still experience the joys of smoked Copper River salmon and lemon-grilled swordfish. (Judith's idiosyncratic definition of the term "animal" is not unusual among vegetarians.)

This twisted moral taxonomy worked fine until Judith ran into Joseph Weldon, who was a graduate student in evolutionary biology. When they first met, Joseph, himself a meat-eater, tried to convince Judith that there is not a shred of moral difference between eating a Cornish hen and a Chilean sea bass. After all, he reasoned, both birds and fish are vertebrates, have brains, and lead social lives. But despite his best efforts, he failed to convince her that, from a culinary ethics perspective, a cod is a chicken is a cow.

Fortunately, their disagreement over the moral status of mahi mahi did not prevent them from falling in love. They married, and her new husband kept the fish-versus-fowl discussion going over the dinner table. After three years of philosophical to-and-fro, Judith sighed one evening and gave in: "OK, I see your point. Fish are animals."

But now she faced a difficult decision: She could either quit eating fish or stop thinking of herself as a vegetarian. Something had to give. A week later, friends invited Joseph to a grouse hunt. Though he had no experience with a shotgun, he somehow managed to hit a bird on the fly, and, in the grand Paleolithic tradition, showed up at home, carcass in hand. Joseph then proceeded to pluck and cook the grouse, which he proudly served to his wife for dinner along with wild rice and a lovely raspberry sauce.

In an instant, fifteen years of moral high ground went down the drain. ("I am a sucker for raspberries," Judith told me.) The taste of roasted grouse opened the floodgates, and there was no going back. Within a week, she was chowing down on cheeseburgers. Judith had joined the ranks of ex-vegetarians, a club which outnumbers current vegetarians in the United States by a ratio of four to one.

■ ■ ■

Then there was Jim Thompson who, when I met him, was a twenty-five-year-old doctoral student in mathematics. Before beginning graduate school, Jim had worked in a poultry research laboratory in Lexington, Kentucky, where one of his jobs consisted of dispatching baby chicks at the end of the experiments. For a while, this posed no problem for Jim. However, things changed one day when he was looking for a magazine to read on a plane and his mother handed him a copy of *The Animals' Agenda*, an animal rights magazine. He never ate meat again.

But that was just the start. Over the next couple of months, Jim stopped wearing leather shoes, and he pressured his girlfriend to go veg. He even began to question the morality of keeping pets, including his beloved white cockatiel. One afternoon Jim looked at the bird flitting around her cage in his living room, and a little voice in his head whispered, "This is wrong." Gently, he carried the bird into his backyard. He said good-bye and released the cockatiel into the gray skies of Raleigh, North Carolina. "It was a great feeling," he told me. "Amazing!" But then he sheepishly added, "I knew she wouldn't survive, that she probably starved. I guess I was doing it for myself more than for her."

■ ■ ■

Our relationships with animals can also be emotionally complicated. Just ask Carol Audette. She fell head over heels for an eleven-hundred-pound manatee. A couple of decades ago, Carol applied for a job—any job—at a small natural history museum in Bradenton, Florida. The museum had an opening; they were looking for a caregiver for Snooty, a thirty-year-old sea cow. Carol had no experience working with marine mammals, but they offered her the position anyway. She did not know that her life was about to change.

On the phylogenetic scale, Snooty fell somewhere between the Creature from the Black Lagoon and Yoda. When Carol introduced me to him, Snooty hooked his flippers over the edge of his pool, hoisted his head two feet out of the water, and looked me straight in the eye, checking me out. While his brain was smaller than a softball, he seemed oddly wise. I found the experience unnerving. Not Carol. She was in love with him.

For the next twenty years, Carol's life revolved around Snooty. She spent nearly every day with him, even coming around to visit on her days off. Food was a major part of their relationship. Manatees are vegetarians, and Carol fed him by hand—120 pounds of leafy green vegetables, mostly lettuce, every day.

But life with an aging sea cow has its downsides. Snooty adored Carol as much as she doted on him. When she and her husband would sneak off for a week or two of vacation, Snooty would get in a funk and quit eating. All too often, Carol would get a call saying that Snooty was off his feed—yet again—and she would rush back to Bradenton to gently ply him with a couple of bushels of iceberg lettuce.

At some point, Carol gave up vacations. That's when her husband accused her of having her priorities screwed up, of loving a half-ton blob of blubber and muscle more than she loved him.

IS IT WRONG TO FEED KITTENS TO BOA CONSTRICTORS?

As a research psychologist, I have been studying human-animal relationships for over thirty years, and I have learned that the quirky thinking about other species we see in Judith, Jim, and Carol is not the exception but the rule.

I began to think seriously about the inconsistencies in our relationships with other species of animals one sunny September morning when I got a phone call from my friend Sandy. At the time, I was an animal behaviorist, and Sandy was a professor at my university. She was also an animal rights activist.

"Hal, I heard that you were picking up kittens from the Jackson County animal shelter and feeding them to a snake. Is it true?"

I was completely taken aback.

"*Arrgh.* What are you talking about? We do have a pet boa constrictor, but he is just a baby. He could not possibly swallow a kitten. And I like cats. Even if he were bigger, I would NEVER let him eat a cat."

Sandy apologized profusely. She said she figured the charge was not true, but she just had to check. I told her that I understood, but that I would appreciate it if she would assure her animal protection pals that I was not dipping into our community's reservoir of unwanted cats to feed my son's snake.

But the incident got me thinking about the moral implications of keeping a predator for a pet. We had acquired the little boa by accident. I was spending the summer as a visiting scientist at the University of Tennessee developing a method for measuring personality differences in snakes. I was in the lab testing animals one day when the phone rang. It was a stressed-out man who had awakened to find that his seven-foot red-tail boa constrictor had given birth during the night to forty-two wriggling newborns. He and his wife were understandably shaken; the new mom had never shown any erotic interest in the male with whom she had shared a cage in the couple's living room for the previous eight years.

The man had heard that I was a snake behaviorist. He wanted tips on how to keep the babies healthy and where he could find good homes for them. I recommended that he contact a reptile expert I knew at the university's veterinary college for information on raising newborn snakes, and I agreed to adopt one of the babies myself. That evening, my eleven-year-old son, Adam, and I drove to the couple's house where they gave him the pick of a very large litter. Adam selected the cutest one and named him Sam.

Sam was a low-maintenance pet. He did not scratch the furniture, keep the neighbors awake, or require daily exercise. He was gentle—except for the time he tried to swallow Adam's thumb. It was Adam's fault. He made the mistake of lifting Sam out of his cage immediately after he had handled his friend's pet hamster. Sam's brain was about as big as an aspirin tablet, and he could not tell the difference between a rodent and a human hand. He just smelled meat.

The accusation that we were feeding kittens to Sam was, of course, ridiculous. While boa constrictors are equal opportunity eaters when it comes to small mammals, Sam was only eighteen inches long and could barely swallow a mouse. Over the next couple of days, however, several questions kept nagging me. My accuser had inadvertently forced me to confront questions about the moral burdens of bringing animals into our lives that I had never really considered. Snakes don't eat carrots or asparagus. So, given Sam's need for meat, was it ethical for us to keep a boa constrictor for a pet? Is having a pet that gets its daily ration of meat from a can of cat food morally preferable to living with a snake? And, finally, are there circumstances in which feeding kittens to boa constrictors might be morally acceptable?

The person who started the rumor about me lived with several cats which she allowed to roam the woods around her house. Like many cat lovers, she conveniently ignored the fact that from lions to tabbies, all members of the family *Felidae* eat flesh for a living. Each day the cats of America chow down on a wide array of meat. The pet-food shelves of my local supermarket are piled high with six-ounce tins of cow, sheep, chicken, horse, turkey, and fish. Even dried cat foods are advertised as containing "fresh meat." With eighty million cats in America, the numbers add up. If each cat consumes two ounces of meat daily, collectively, they would eat nearly ten million pounds of flesh—the equivalent of two million chickens—every single day.

Further, unlike snakes, cats are recreational killers. In the United States alone, it has been estimated that between seven and twenty-five billion small animals a year fall victim to the claws of pet cats. Oddly, most cat owners don't seem to care about the devastation their feline friends wreak on wildlife. Kansas investigators informed a group of cat owners about the devastating effects of cats on local songbird populations. Then the researchers asked if they would now keep their cats indoors. Three-fourths of the respondents said no. In a cruel irony, many cat owners also enjoy feeding birds in their backyards, inadvertently luring legions of hapless towhees and cardinals to their deaths at the claws of the family pet.

So, pet cats cause havoc. What about pet snakes? Well, first, there are a lot fewer of them. Likely, there are at least a hundred times more cats living in American homes than there are snakes. Also, each snake consumes only a fraction of the flesh that a cat does. According to Harry Greene, a Cornell University herpetologist who studies the feeding ecology of tropical snakes, an adult boa living in a Costa Rican rainforest consumes maybe half a dozen rats a year. This means that a medium-size pet boa constrictor needs less than five pounds of meat a year to stay in good condition. A pet cat requires far more flesh. At two ounces a day, the average cat would consume about fifty pounds of meat a year. It can be argued that the moral burden of enjoying the company of a cat is ten times higher than that of enjoying the company of a snake.

Nearly a million unwanted cats are euthanized in animal "shelters" in the United States each year. Their bodies are typically cremated. Wouldn't it make sense to make these carcasses available to snake fanciers? After all, these cats are going to die anyway, and fewer mice and rats would be sacrificed to satisfy the dietary needs of the pet boas and king snakes living in American homes. Seems like a win-win, right?

Yikes . . . I had inadvertently painted myself into a logical corner in which feeding the bodies of dead kittens to boa constrictors would not only be permissible, but morally preferable, to feeding them mice. But while the logical part of my brain may have concluded that there was not much difference between raising snakes on a diet of mice or kittens, the emotional part of me was not buying the argument at all. I found the idea of feeding the bodies of pet cats to snakes revolting, and I had no intention of hitting up the county animal shelter for kitten carcasses.

THE PARADOXES OF PET-KEEPING

The boa constrictor incident got me thinking about other instances of morally problematic interactions between people and animals that I had encountered. For instance, my graduate school friend Ron Neibor's doctoral dissertation was on how the brain reorganizes itself after injury. This was before researchers used imaging technologies such as MRIs to study changes in the brain. Cats, unfortunately, were the best model for the neural mechanisms he was studying. He used a standard neuroscience technique at the time, surgically destroying parts of the animals' brains and then observing how they recovered over the succeeding months. The problem was that Ron came to like his cats. His study lasted a year, during which time he became attached to the two dozen animals in his lab. On weekends, he would drive to the lab, release the kitties from their cages, and play with them on the floor of the animal colony. They had become pets.

His experimental protocol required that he confirm the location of the neurological lesions in the animals in the experimental group by examining their brain tissue. Part of this procedure—technically referred to as perfusion—is not pleasant under the best of circumstances. Each animal had to be injected with a lethal dose of anesthetic. Then, formalin was pumped through its veins to harden the brain, and the animal's head was severed from the body. Pliers were used to chip away the skull so the brain could be extracted and sliced into thin sections for microscopic analysis.

It took Ron several weeks to perfuse all the cats. His personality changed. A naturally cheerful and warm-hearted person, he became tense, withdrawn, shaky. Several of the graduate students in his lab became concerned and offered to perfuse his cats for him. Ron refused. He was unwilling to dodge the moral consequences of his research. He did not talk much during the weeks he was "sacrificing" his cats. Killing them took a toll. Sometimes Ron's eyes were red, and he looked down as he passed me in the halls.

These sorts of moral complexities also extend to man's best friend, the dog. My neighbor Sammy Hensley, a farmer who lived just down the road from us when we lived on Sugar Creek in Barnardsville, North Carolina is an example. His two passions were dogs and raccoon hunting. Coon hunting wasn't so much sport for Sammy as a way of life. He didn't eat the raccoons he killed. He skinned them out and nailed their pelts and paws to the side of his barn so his neighbors could track his success during hunting season. (It was while helping him skin a coon that I learned raccoons—and most mammals—have a bone in their penis; humans are one of the exceptions.) I once accused him of nailing the skins up just to irritate my wife, Mary Jean, who once had a pet raccoon and is nuts about them. But it really wasn't about that. It was just the North Carolina mountain way.

There were two kinds of dogs in Sammy's life—pet dogs and coon hounds—and they led very different lives. He kept four or five hounds at a time—a couple of experienced hunters and a pup or two in training. I loved the names of the breeds—treeing Walkers, Plott hounds, blueticks, redbones. Lanky animals with deep voices, languid eyes, oily coats, and the pungent smell hounds have. They usually looked lethargic. That's because they lived most of their lives lying in the dirt, tethered to dog houses by eight-foot chains. But they came alive during hunting season, when they got to tear through the rhododendron thickets in the middle of the night, baying, nose to the ground. You could hear their voices all through our cove.

Sammy loved his hounds. He could tell their voices apart; he knew by the tenor of their yips and yells when they had treed the coon (good) or when they were on a possum's trail (not good). He worried when they got lost and didn't come home in the morning. But they were working dogs, not pets. If a dog couldn't do its job, Sammy would sell it or swap it for a new one.

But Sammy and his wife, Betty Sue, also had pet dogs. While the hounds never saw the inside of their house, the pets—smallish animals like beagles and Boston terriers—had the run of the place. Unlike the coon hounds, these dogs were part of the family. They were petted and played with and allowed to beg for food at the dinner table.

One afternoon, when Sammy was mowing hay on a steep section of a hillside pasture, his tractor flipped over, killing him. After Sammy died, Betty Sue didn't keep the hounds long, but their little Boston terrier helped her get through the tough times. In the Hensley home, the hounds and the pet dogs might as well have been different species.

Most of the dogs living in American homes are simply companions, but our attitudes toward them can be as convoluted as Sammy's relationships with the two categories of dogs in his life. Surveys have found that between 80% and 90% of dog owners think of their pets as family members. Yet there is a dark side to our interactions with dogs. One in ten American adults is afraid of dogs, and dogs are second only to late-night noise as a source of conflict between neighbors. (My friend Ross had to sell his house because his neighbor's barking dogs turned his life into a nightmare.) In a typical year, four and a half million Americans are bitten by dogs, and three dozen people, mostly children, are killed by them.

The dog's-eye view of the human-pet relationship isn't always rosy either. In the United States, nearly seven hundred thousand dogs are euthanized in animal shelters each year. Then there are the horrendous genetic problems we have inflicted upon dogs in our attempts to breed the perfect pet. Take, for example, the English bulldog, a breed that anthrozoologist James Serpell refers to as a canine train wreck. Bulldogs have such monstrous heads that 90% of bulldog puppies have to be delivered by Caesarian section. Their distorted snouts and deformed nasal passages make breathing a chore, and they suffer from joint diseases, chronic dental problems, deafness, and a host of dermatological conditions caused by their wrinkled skin. To add insult to injury, bulldogs also easily overheat and tend to slobber, snore, fart, and suddenly drop dead from cardiac arrest.

Things are worse for dogs in parts of Asia, where a puppy can be a pet or an item on the menu. While dogmeat costs more than chicken (about seven dollars a pound in Seoul), South Koreans eat about a million dogs every year. Meat dogs, which are typically short-haired medium-size animals that look disconcertingly like Old Yeller, are raised in horrific conditions before they are slaughtered, usually by electrocution.

We tend to ignore these contradictions but as a psychologist, they fascinated me.

FROM THE BEHAVIOR OF ANIMALS TO THE BEHAVIOR OF ANIMAL PEOPLE

In the weeks after I was accused of feeding kittens to Sam, I found myself thinking more about the paradoxes associated with our relationships with animals and less about my animal behavior studies. By conventional professional standards, my animal behavior research program was going well. I was publishing articles in good journals, receiving my share of grant funding, and presenting my research at international conferences. But it dawned on me that there were plenty of smart young scientists investigating topics like vocalizations in cotton rats, tool use in crows, and the offbeat reproductive habits of spotted hyenas (female hyenas give birth through their penises). On the other hand, there were only a handful of researchers trying to understand the often wacky ways that people relate to other species. It was an emerging field, one that I could enter on the ground floor and make a contribution. Within a year, I had closed up my animal lab and was concentrating full-tilt on the psychology of human-animal interactions.

Since shifting from studying the behavior of snakes and alligators to studying animal people, much of my research has focused on individuals who love animals but who confront moral quandaries in their relationships with them—the veterinary student who tries not to cry when she euthanizes a puppy, a woman whose marriage broke up because her husband objected to her growing involvement in the animal rights movement, the burly circus animal trainer whose life is completely focused on the giant brown bears he hauls around the country in the dreary confines of an eighteen-wheeler, the grizzled cockfighter who beams when I offer to take a picture of his beloved, battle-scarred, seven-time winner.

I have attended animal rights protests, serpent-handling church services, and clandestine rooster fights. I have interviewed laboratory animal technicians, big-time professional dog show handlers, and small-time circus animal trainers. I've watched high school kids dissect their first fetal pigs and helped a farm crew slaughter cattle. My students and I have analyzed several thousand Internet bulletin board messages between biomedical researchers and animal rights activists as they tried—and ultimately failed—to find common ground. We have studied women hunters, dog rescuers, ex-vegetarians, and people who love pet rats. We have surveyed thousands of people about their attitudes toward rodeos, factory farming, and animal research. We even poured over hundreds of back issues of sleazy supermarket tabloids, analyzing modern cultural myths about animals. (The original title of our article on tabloid animal stories was "Woman Gives Birth to Litter of Nine Rabbits." Unfortunately, the editor of the journal to which we submitted the manuscript did not find the title to be sufficiently scientific and insisted we change it.)

Like most people, I am conflicted about our ethical obligations to animals. The philosopher Strachan Donnelley named this murky ethical territory the "troubled middle." Those of us in the troubled middle live in a complex moral universe. I eat meat, but not as much as I used to, and not veal. I oppose testing the toxicity of oven cleaner and eye shadow on animals, but I would sacrifice a lot of mice to find a cure for cancer. And while I find some of the logic of animal liberation philosophers convincing, I also believe that our vastly greater capacity for symbolic language, culture, and ethical judgment puts humans on a different moral plane from that of other animals. We middlers see the world in shades of gray rather than the clear blacks and whites of committed animal activists and their equally vociferous opponents. Some argue that we are fence-sitters, moral wimps. I believe, however, that the troubled middle makes perfect sense because moral quagmires are inevitable in a species with a huge brain and a big heart. They come with the territory.

I wrote *Some We Love, Some We Hate, Some We Eat* for anyone interested in human-animal relationships. As a researcher, I normally write for specialists whose job it is to wade through boring jargon-laden prose that can quickly make your eyes glaze over. But I am convinced that scientists should communicate with the public, people who do not know the difference between an analysis of variance and a factor analysis but who are eager to read about current research findings and the hot controversies in our field. The trick is to inform readers about the latest results in an interesting way, but at the same time respect the complexity of the issues and be honest about what we know and what we don't.

Many of the topics in the book are controversial. Researchers disagree about the effectiveness of animal-assisted therapies, whether children who abuse animals usually become violent adults, and why most vegetarians and vegans go back to eating meat. The passions of the public run high over contentious issues such as banning dog breeds like pit bulls or using monkeys to discover a vaccine that could help stop a global pandemic. Some of these debates have become bitterly divisive, with the partisans approaching the issues with a passion approaching religious zeal.

For the most part, I have tried to approach these issues as objectively as I can. This means, of course, that well-intended and intelligent people on both sides of some of these controversies will sometimes disagree with me. To this end, I have included an extensive list of research citations and recommended readings at the end of the book. If you want to delve further into the research on the effects of pets on human health or the psychology of animal activism, I point you to relevant studies. My goal is not to change your mind about how we should treat animals. Rather, I want to encourage you to think more deeply about the psychological and moral implications of some of our most important relationships, the connections we have with the nonhuman creatures in our lives.

Late one afternoon in 1986, I was standing in a hallway of a posh Boston hotel deep in conversation with Andrew Rowan, who was the director of the Center for Animals and Public Policy at Tufts University. He had organized one of the first international conferences on human-animal relationships, and we were discussing the paradoxes that so often crop up in our attitudes toward the use of animals. Why, for example, can dogs eat at the dinner table in some countries, but be eaten for dinner in others? How could Adolf Hitler have been responsible for enacting the world's most progressive animal welfare legislation? How can 60% of Americans tell pollsters that animals have the right to live while over 95% of Americans regularly consume them?

Andrew looked up at me and said, "The only consistency in the way humans think about animals is inconsistency."

This book is my attempt to explain this paradox.

1

Anthrozoology

THE NEW SCIENCE OF HUMAN-ANIMAL INTERACTIONS

> Our failure to study our relationships with other animals has
> occurred for many reasons. Much of it can be boiled down
> to two rather unattractive human qualities: arrogance and
> ignorance.
>
> —CLIFTON FLYNN, SOCIOLOGIST

The thirty-minute drive from the Kansas City airport to the conference
hotel was much more interesting than the three-hour flight from North
Carolina. It was the fall of 2009, and I had flown in for the annual meeting of the International Society of Anthrozoology. I found myself sharing
a ride with a woman named Layla Esposito. She was a social psychologist who told me she recently completed her PhD dissertation on bul-
lying among middle school children. Puzzled, I asked her why she was
attending a meeting on the relationships between people and animals.
She said she was a program director at the National Institute of Child
Health and Human Development. She was at the conference to let researchers know about a new federal grant program to fund research on
the effects of animals on human health and well-being. The money was
coming from a public-private partnership between the National Institutes
of Health and Mars, the corporate giant that makes candy bars for me and
crunchy salmon treats for my cat. The NIH was interested in research on
the impact of pets on children and older adults. Is pet therapy an effective

treatment for kids with ADHD? Are elderly people who live with dogs healthier and happier?

"How much money are you giving out?" I asked. "About two million dollars a year," she said. "Fantastic! This is just what the field needs." I was thinking Layla was going to have a full dance card for the next couple of days.

WHY OUR RELATIONSHIPS WITH ANIMALS MATTER

While $2 million seemed a paltry sum compared to the $7 billion the National Cancer Institute doles out, these funds have been a shot in the arm for anthrozoology—the study of our interactions with other species. According to a recent progress report, the NIH/Mars program has funded dozens of studies—research on the impact of therapeutic horseback riding on children with autism, predicting which puppies have the potential to become first-rate service dogs, and how pet cats fit into family systems.

While animals are important in so many aspects of human life, the study of our interactions with other species has been neglected by scientists until recently. Take my field, psychology. For a hundred years, psychologists concentrated their efforts on uncovering behavioral and cognitive processes such as motivation, perception, and memory, but neglected important facets of daily life like eating, religion, and how we spend our leisure time. Our relationships with animals, especially our pets, fall into the category of things that everyday people care about, but psychologists don't.

One reason behavioral scientists shy away from studying human-animal interactions is that, for many of them, the topic seems trivial. This attitude is wrong-headed. Understanding the psychology underlying our attitudes and behaviors toward other species is important for several reasons. Nearly three out of four Americans live with animals, and many people have deep personal relationships with their pets. Also, our beliefs about how we should treat other species are ethically challenging. Many of us, for example, are torn over whether animals should be used as subjects for biomedical research or killed because they taste good. Finally, people are fascinated by anthrozoological research. When I tell someone that I study human-animal interactions, almost inevitably they tell me stories about their wacky dog or their objections to eating meat or how their Aunt Sally loves to hunt bears with her Plott hounds.

THINKING LIKE AN ANTHROZOOLOGIST

Anthrozoology transcends normal academic boundaries. Among our numbers are psychologists, veterinarians, animal behaviorists, historians, sociologists, and anthropologists. As in every science, anthrozoologists don't always see eye to eye. We differ in our attitudes toward some of the thorny moral issues that arise in human-animal relationships. We don't even agree on the name of our discipline—some prefer to call it human-animal studies. Despite these differences, researchers who study our relationships with animals have a lot in common. We all believe that our interactions with other species are an important component of human life and we hope our research makes the lives of animals better.

As academic disciplines go, anthrozoology is a small pond, but in the last three decades, we have come a long way. Several dozen journals are now devoted to publishing our studies, and the International Society for Anthrozoology holds annual meetings where researchers report their latest findings and argue about whether walking your dog will cause you to lose weight and how long cats have been domesticated. Over fifty colleges and universities offer degree programs in human-animal studies, and research centers have been established in North America, Europe, Asia, and Australia.

To get a sense of anthrozoological research, here are a few examples of hot issues in the science of human-animal interactions.

DO DOLPHINS AND DOGS MAKE
GOOD THERAPISTS?

One of the most important issues in anthrozoology is whether interacting with animals can alleviate human suffering. Animal-assisted therapies have been around for decades. Sigmund Freud, the founder of psychoanalysis, found that some of his patients would be more willing to talk about painful issues when Jofi, his Chow Chow, was in the room. The term "pet therapy" was coined in 1964 when Boris Levinson, a child psychologist, reported that some troubled children would open up when they played with his dog Jingles. The residents in my ninety-two-year-old mother's assisted-living facility perked up when the therapy dogs visited a couple of times a week. And I find that spilling my guts out to Tilly helps me work out my little problems. (My cat Tilly takes a tough-love approach to counseling. This means that when I start to whine, she just sniffs and walks away.)

But how good is evidence that animals make good therapists? The most controversial form of animal-assisted therapy involves animals held in captivity against their will—dolphins. Claims made about the curative powers of these marine mammals are over the top. Interacting with dolphins, it is alleged, can alleviate Down syndrome, AIDS, chronic back pain, epilepsy, cerebral palsy, autism, learning disorders, deafness, and can even shrink tumors. Among the proposed healing mechanisms are bioenergy force fields, the high-frequency clicks that dolphins use to communicate with each other, and a presumed ability to alter human brain waves.

Dolphin therapy sounds great. Go swimming, get well. But before you sign up for a couple of weeks in a dolphin tank, you should check out the science behind these claims. Most of these rely on anecdotes, self-reports, or poorly designed experiments conducted by researchers associated with for-profit swim-with-dolphins programs. Unfortunately, dolphin therapy is particularly attractive to desperate parents who will pay whatever it takes to help their kids with disorders such as autism and Down syndrome. They flock in droves to the more than one hundred swim-with-dolphins programs in places like the Florida Keys, Bali, Great Britain, Russia, the Bahamas, Australia, Israel, and Dubai—all of them hoping that, through some unknown force, these creatures with perpetual Mona Lisa smiles will work their magic on broken minds and tormented souls. Dolphin therapy can cost more than $1,000 a day. Is the money well spent? Will their hopes be fulfilled?

Nature does not give up its secrets easily. Scientists have to work hard to get beneath the veil. Researchers, just like everyone else, can be duped, particularly when they have a horse in the race. That's why graduate students take courses in research methods and statistics: to learn the tricks of the trade that keep them honest. We throw around phrases like "internal and external validity," "placebo control," "random assignment," "single- and double-blind experiments," and "correlation is not causality." I won't bore you with the details except to say that these conceptual tools help reduce the chances that we will unconsciously tilt the playing field our way.

Good scientists are perpetually on the lookout for alternative explanations for research results even if they crush our pet ideas. Think about it. In addition to hanging out with some of the most appealing creatures on earth, you travel to beautiful places, spend time floating in tropical seas, and live for a while in a supportive environment where your expectations for success are high. How can researchers separate the impact of interacting with a dolphin from all the other neat things that happen during a week at dolphin camp? Fortunately, there are methods to help tease out the real effects of treatments from those caused by unconscious biases that can creep into our experiments.

To take a cold, hard look at whether the benefits of interacting with dolphins are due to more than just temporary feel-good emotions, we need to use a *Consumer Reports*-type approach. What, for example, is the effect of ultrahigh-frequency dolphin sounds on handicapped children? German researchers took this question on. They carefully observed sessions in which dolphins interacted with mentally and physically handicapped kids who were in a Florida dolphin therapy program. They found that most of the dolphins ignored the children, and there was not much ultrasonic dolphin talk going on. Indeed, the children were exposed to an average of only ten seconds of dolphin ultrasounds during each session, not nearly enough to be beneficial. The researchers concluded that the kids would have been better off playing with dogs.

But what about the dolphins' purported ability to heal through good vibrations, a healing smile, and mysterious electric fields? Careful analyses of these claims have been conducted by several groups of scientists. Among the most well-known are Lori Marino and the late Scott Lilienfeld. Lori was the first scientist to demonstrate that dolphins can recognize themselves in mirrors (a trait they share with humans, apes, and possibly elephants and cleaner fish). Scott was a clinical psychologist who made a career out of taking on some of psychology's most sacred cows, such as whether those Rorschach inkblots reveal much about your personality (they don't).

Given Lori's expertise with dolphins and Scott's ability to cut through psychobabble, they were the perfect team to assess whether dolphin therapy has a demonstrable effect on troubled bodies and minds. In a series of research papers, Lori and Scott carefully evaluated the methods of each of the thirteen published studies claiming that dolphin therapy is effective for mental disorders—depression, developmental disabilities, autism, and anxiety. They found that all of them were methodologically flawed. The problems included small sample sizes, lack of objective measures of improvement, and inadequate control groups. Then there was the inability to separate the effects of the dolphins from the feelings of well-being that come from doing new things in pleasant environments, and conflicts of interests on the part of the investigators.

Lori believes there is no valid scientific evidence that dolphin therapy is an effective treatment for any of the disorders that its advocates claim. She says it is all pseudoscience. Not content with blowing off dolphin therapy as scientific mumbo jumbo, Lori wants to put the industry out of business. She calls it a dangerous fad. I can see the fad part, but why is it dangerous? If you can afford it, why not let kids with too little joy in their lives frolic with Flipper for a couple of weeks? Seems harmless.

Lori doesn't agree. She points out that this form of animal-assisted therapy poses risks for both humans and animals. Dolphins can be aggressive, even to the kids they are supposed to be healing. One study found that half of four hundred people who worked professionally with marine mammals had suffered traumatic injuries, and participants in dolphin therapy programs have been slapped, bitten, rammed (the latter resulting in a broken rib and a punctured lung). You can even contract skin diseases from these "therapists."

Dolphin therapy also raises pesky ethical issues. Clinical psychologists choose to become therapists. Dolphins do not. Do we have the right to capture intelligent animals with complex social lives and sophisticated communication systems and turn them into therapists for our damaged children? I suppose the practice might be justified if these animals did possess special curative powers. But I would need rock-solid evidence that dolphins could break through the isolation of a child with autism or that dolphin-generated bioelectric force fields could jolt the middle-age depressive out of his debilitating funk. Unfortunately, that evidence is about as solid as the support for back-to-the-womb regression therapy.

Clearly, we should not be forcing captive dolphins to serve as surrogate psychotherapists. But what about other forms of animal-assisted therapy—say interacting with a trained therapy dog or therapeutic horseback riding? In 2007, researchers from the University of Utah analyzed the results of forty-nine studies on the effectiveness of animal-assisted therapies. They reported that in nearly all of the studies, the subjects accrued benefits from interacting with their nonhuman therapists. The problem was that, like dolphin therapy research, most of these studies were seriously flawed.

Recent better-designed studies on psychological interventions involving animals have produced mixed results. In some cases, the results are encouraging. Investigators from Purdue University's Center for the Human-Animal Bond found that military veterans suffering from post-traumatic stress disorder who received a psychiatric service dog were less depressed and had a higher quality of life and more friends than vets on a service dog waiting list. And in a well-designed study, researchers at the University of British Columbia found that just twenty minutes of interacting with a therapy dog reduced homesickness in college students.

Some of the results of the newer studies, however, have been disappointing. Perhaps the most ambitious investigation of the effectiveness of animal-assisted therapy was a randomized controlled trial involving kids undergoing outpatient treatment for cancer. The research was funded by Zoetis, a veterinary pharmaceutical company. The research took seven years from start to finish and cost $1 million. It involved 106 children and teens, their parents, and twenty-six therapy dog teams. The participants in the treatment group received visits from the therapy dogs once a week for four months, while the control group got standard treatment.

The researchers expected the kids in the dog group would be less stressed and anxious than children in the no-dog group and would score higher on a series of quality of life measures. Unfortunately, they were wrong. The patients in the treatment group loved playing with the therapy dogs. But at the end of the study, they were no less anxious, stressed, or happier than children in the control group.

Other researchers have found similarly disappointing results. For example, a well-designed study by Megan Mueller and her colleagues at Tufts University found that therapy dogs had zero effect when it came to reducing anxiety in acutely stressed-out teens. Danish researchers examined the impact of interacting with a therapy dog or a cute robotic seal on nursing home residents. The study lasted six weeks, and the initial results were encouraging. But after six weeks, the therapy dog did not have any impact on the health or well-being of the participants.

In recent years there has been an explosion of research on animal-assisted therapies. Between 2000 and 2020, the annual number of published papers on the use of therapeutic dogs, birds, guinea pigs, and even crickets increased tenfold. Unfortunately, most of these studies are seriously flawed. For example, researchers from Baylor University found that the results of 90% of studies on the impact of interacting with animals on kids with autism could not be trusted. The most common problems with these studies are no control groups, too few subjects, and separating the impact of the therapy animals from the benefits of talking with their sympathetic handlers. The hopes and expectations of the investigators can also affect research results. People like me and my fellow anthrozoologists are attracted to the field because we are passionate about animals. We want to believe that petting therapy dogs or a treatment course involving therapeutic horseback riding will heal damaged minds. The problem is that *a priori* convictions do not necessarily go hand in hand with good science.

Because of bestselling books with titles like *Doctor Dogs: How Our Best Friends Are Becoming Our Best Medicine* and newspaper headlines like "National Study Shows Therapy Dogs Can Aid Kids Undergoing Cancer Treatment," the idea that animals, particularly dogs, make good therapists has become widely accepted. I'm certain that, in some cases, some animal therapies work for some individuals some of the time. But I'm equally sure that there is a mismatch between what most people have come to believe about the curative powers of interacting with animals and the more nuanced pattern of actual research results.

WILL GIVING UP MEAT AFFECT YOUR LOVE LIFE?

When I became serious about studying human-animal interactions, one of my first research projects focused on how people became involved in the animal rights movement and how it affected their lives. One interview was with a woman I will call Nancy. Like most serious animal activists, she was a vegetarian. But her husband, a Duke University professor, loved to eat meat. I asked Nancy if their dietary differences affected their relationship. "Not at all," she replied. "I will buy meat for him and I even cook it for him." But then she added, "But, of course, I would not kiss him after he ate meat."

Despite not kissing her husband after dinner, Nancy had made an easy accommodation to his dietary preferences. But for other animal activists, things did not go so smoothly. Take Elizabeth. She told me, "Vegetarianism definitely interferes with my social life. I won't go out with anyone who is not a vegetarian. It limits the pool of possible men. Early on, most of the men I dated were not vegetarian. I will never do that again. Having that kind of moral blockade between someone you are involved with is just impossible."

When it comes to friendships, love, and romance, are most vegetarians and vegans like Nancy, whose husband's penchant for meat was not a big deal? Or are they more like Elizabeth, who never dates meat-eaters? With only a few exceptions, nearly all of my married vegetarian friends have vegetarian spouses. But does empirical research show that going vegetarian will change our choices in friends and lovers? The answer is yes, according to research by John Nezlek and his colleagues.

About 2,500 subjects participated in a series of four studies. They included representative samples of adults in the United States and Poland as well as American college students. In the studies, individuals were asked to describe their diets. In the first study, American adults also completed questionnaires related to the importance of the food they ate on their social identities and sense of self. The other studies examined how the choice of diet affected whom the participants chose as best friends and romantic partners.

As the researchers predicted, vegetarians and vegans were much more likely than omnivores to say that food was important in shaping their self-identity. Further, diets were a major influence on close friendships. Among the Americans in the studies, vegetarians and vegans were *three times* more likely than the omnivores to indicate their closest friends were nonmeat-eaters. The difference among the Polish participants was even greater: In that group, vegetarians/vegans were *six times* as likely as omnivores to have best buds that did not eat animals. But the most interesting finding was related to lovers. Vegetarians and vegans were twelve times more likely than omnivores to have romantic partners who did not consume meat.

These results raise important issues. Many animal activists have told me their involvement in animal protection had resulted in alienation from their families and friends. I suspect a lot of this was due to their dietary changes. Across all cultures, a great deal of human social life revolves around the sharing of food and drink. It's safe to say that a moral vegetarian would not feel comfortable sitting down to a family Christmas dinner in which the main course was a five-pound hunk of rare rib roast. Similarly, as an older woman in a class I was teaching recently told me, many of her longtime friends stopped asking her over to their homes for dinner soon after she became a vegan. They just did not know how to cook for her.

This research has implications for the love lives of vegetarians and vegans. According to the study, giving up meat can shrink your dating pool considerably. And I suspect this is a big problem for ethical vegans. According to a 2018 Gallup poll, only 3% of Americans don't consume any animal-related products. When I suggested to one of the authors of the study, Catherine Forestell, director of neuroscience and associate professor of psychological sciences at William & Mary, that this could result in a 97% reduction in an American vegan's potential dating partners, she agreed—but then cautioned me in an email:

Those numbers probably depend to some extent on age and geographical region. I expect that a 20- to 30-year-old vegan who lives in Colorado would have an easier time finding a like-minded mate than a 60- to 70-year-old vegan in Alabama.

The romantic angle of the research hits home with me. My son Adam and his wife have now been together for twenty years. His wife gave up meat in her teens, but Adam is an inveterate carnivore whose freezer is usually filled with beef and pork he buys from a local farm. At this point, our five- and ten-year-old grandsons are omnivores. But it will be interesting to see how being raised in a culinary mixed marriage will affect their future dietary choices.

HOW OUR PERSONALITY AFFECTS
OUR CHOICES IN PETS

My friends Phyllis and Bill have a mixed marriage. She is a cat person; he is not. Phyllis has had cats since she was in college, usually two or three at a time. I once spent a month house-sitting for her and agreed to give one of her cats, the foul-tempered Chris, two pills every day, one for epilepsy, the other for depression. It was a struggle I won only part of the time. Over the years, Phyllis forked out thousands of dollars to veterinarians for patching up Chipper, her gray tabby, who had a penchant for going into full battle with stray tomcats and raccoons.

What does Phyllis like about cats? She claims it's their nicely balanced needs for both affection and independence, a mix she also likes in a husband. She thinks dogs are suck-ups.

Bill, on the other hand, does not particularly like animals. He never has. His parents didn't have pets when he was growing up, and Bill never felt the slightest desire to live with one himself. But then he married Phyllis and he was suddenly living with cats. Over time, his attitudes toward the cats in his home have shifted a bit, from indifference to tolerance. He admits that he enjoys letting one of them lie on his belly every night when he watches the news. But he doesn't feed the cats and he never asks how they are doing when he talks to Phyllis on the phone when he is away. Bill says that if he were living by himself, he would not have a pet at all.

Phyllis is a psychotherapist, a good one. Given her clinical expertise, I asked if she saw a difference between cat people and dog people. I was surprised when she said no, that it was not personality but serendipity that determines the types of pets people fall for. A cute kitten simply shows up in your backyard or you happen to grow up in a family that has dogs or you want an animal companion that will get rid of the mice in your basement.

I am almost certain that you think of yourself as either a dog person or a cat person, probably a dog person. That's because, if asked, most people will instantly put themselves into one of these categories. And according to a recent Gallup poll, 70% of Americans say they are dog people. This aspect of pet demography is paradoxical as there are more cats than dogs in American homes. (Mary Jean and I, by the way, consider ourselves dog people even though we live with a cat.)

But is it true that dog people and cat people have different personalities, or is this yet another piece of common sense that proves to be wrong?

This question was taken on by Sam Gosling, a psychologist at the University of Texas who studies individual differences in people and in animals. His research has shown that while some of our personal preferences reveal aspects of our personality traits, others do not. He can, for example, tell a lot about you from knowing what music you download, how messy your bedroom is, and whether you hang inspirational posters on your wall. On the other hand, he has found that knowing the contents of your refrigerator says nothing about what you really are like.

But before we can take on the question of dog people versus cat people, a brief lesson in the psychology of personality is in order. Psychologists have been arguing about the nature of human personality for a hundred years. One issue they fight about is how many personality traits there are. While there are a few holdouts, most psychologists agree that we can get a good description of a person's personality by measuring five basic traits. (Technically this is referred to as the Five Factor Model; psychologists just call it the Big Five.)

The Big Five traits are:

- Openness vs. Closed to Experience
- Conscientiousness vs. Impulsiveness
- Extroversion vs. Introversion
- Agreeableness vs. Antagonism
- Neuroticism vs. Emotional Stability

Sam and the anthrozoologist Anthony Podberscek wondered if the personalities of pet owners were different than non-pet owners. They scoured the scientific literature, locating dozens of studies comparing the two groups, and found a hodgepodge of results. For every study reporting that pet owners were more extroverted or more emotionally stable or less independent than non-pet owners, there was another one that found no difference between the two groups. They concluded that there was no evidence that pet owners were different than non-pet owners in their basic personalities.

Is this also true of the dog person/cat person dichotomy? Sam maintains an online version of the Big Five Personality Test that tens of thousands of people have taken. Once, for a week, he temporarily added an item in which participants were asked if they considered themselves to be a dog person, a cat person, neither, or both. To his surprise, 2,088 dog people and 527 cat people took the personality test.

Here are their results:

- Dog people are more *extroverted*.
- Dog people are more *agreeable*.
- Dog people are more *conscientious*.
- Cat people are more *anxious*.
- Cat people are more *open to new experiences*.

Subsequent research has generally replicated these results, particularly the finding that cat people are more anxious than dog people. So, there seems to be some truth to the conventional wisdom that dog people and cat people are different. Most of the differences in personalities fall along lines you probably would have predicted. But in science, there is often a catch. In this case, the catch is that the differences in their personality scores were relatively small. The bottom line is that whether you call yourself a dog or a cat person tells us something about your personality— not as much as your music downloads but more than the contents of your refrigerator.

These studies explored personality differences between self-identified cat people and dog people. But anthrozoologists have also examined how personality differences play out in our interactions with companion animals. Take, for example, the Big Five Neuroticism/Anxiety Scale. A slew of studies has found that dog owners who score high on this trait are more likely to have problems with their pets. Carri Westgarth at the University of Liverpool reported that people who scored high on the Neuroticism Scale were 22% more likely to have been bitten by a dog. Similarly, Slovenian researchers found that dogs who were aggressive toward people tended to have owners with high neuroticism scores. Research teams from Hungry and Australia independently gave dog owners the standard Big Five Inventory and their dogs a canine version of the test. Both teams reported that neurotic owners tended to have neurotic dogs. And University of Pennsylvania researchers found that anxious owners more often had fearful dogs who would try to bite their owners.

There seems to be a pattern here. But why should neurotic owners tend to have neurotic and/or aggressive dogs? Several possibilities come to mind. The first is that it is simply a matter of perception. That is, neurotic owners project their anxieties onto their perfectly normal pets. The Hungarian researchers, however, controlled this possibility by having outside observers in their study rate the behaviors of the dogs.

A more plausible explanation is that neurotic owners somehow cause their dogs to be more anxious. This explanation was suggested by investigators from the University of Vienna. In addition to measuring neuroticism in dogs and their owners, they also looked at levels of cortisol, a stress hormone. They found that owners high in neuroticism tended to have dogs with low cortisol variability, a sign of anxiety. And studies have found that dogs are sensitive to the emotions of their owners. The researchers raise the intriguing possibility that the link between neuroticism in owners and their dogs could be the result of emotional contagion.

But the causal arrow might point in a different direction. Social psychologists have found that people tend to have friends who are similar to themselves. This same process could be at work when we choose our companion animals. That is, tense, anxious people might be attracted to breeds of dogs with similar personalities.

As scientists so often say, "More research is needed."

DO CHILDREN WHO ABUSE ANIMALS BECOME VIOLENT ADULTS?

I once spent an afternoon strolling through New York's Metropolitan Museum of Art looking for paintings depicting human-animal relationships. There were lots of them, but one of the most striking was a small oil painting by a sixteenth-century Italian artist named Annibale Carracci aptly titled *Two Children Teasing a Cat*. The painting depicts an innocent-looking young boy and girl and a cat. The boy is holding the cat with his left hand and has a large crayfish in his right. He has provoked the crayfish into clamping one of its massive claws onto the cat's ear. What makes the painting especially chilling is that the children are smiling, delighted with their "game." What should we make of this wanton cruelty? Is it just a childish prank or an indicator of deep-seated psychopathology that will someday erupt into far worse violence?

The infliction of abject cruelty toward members of other species illustrates how our interactions with animals reflect larger themes in psychology. Some scientists believe the roots of cruelty lie in our evolutionary history, particularly the fact that our ancestors were meat-eating apes that delighted in ripping their prey to pieces. Others, however, argue that human children are naturally kind. They argue that callousness toward animals is instilled by a culture that promotes activities like hunting and the consumption of flesh. Cruelty also offers fodder for those looking for moral inconsistencies in our treatment of animals. What, for instance, is the moral difference between the pleasure that a hunter derives from shooting deer and that a child gets from tying a tin can to the tail of a dog?

The anthropologist Margaret Mead wrote that "one of the most dangerous things that can happen to a child is to kill or torture an animal and get away with it." She was echoing a theme that has been knocking around for hundreds of years. The Enlightenment philosophers John Locke and Immanuel Kant connected animal cruelty and human-directed violence. Indeed, Kant argued that the only reason we should be nice to animals is that animal cruelty leads to human brutality. Some anthrozoologists are convinced that animal abuse in children is often the first step on a path that leads to adult criminality. Others, however, are not so sure.

One of the first systematic studies associating animal cruelty and criminality was conducted by Alan Felthous and Stephen Kellert. They interviewed groups of aggressive criminals, non-aggressive criminals, and non-criminals. The highly aggressive criminals in their study were five times more likely to repeatedly abuse animals than men in the other groups. And their level of violence was different. They cooked live cats in microwave ovens, drowned dogs, and tortured frogs.

In the wake of this and similar studies, I began asking my friends if they ever abused animals when they were children. It was an eye-opener, particularly since many of them are successful college professors. For example, my buddy Fred confessed that he and his childhood pals blew up frogs with firecrackers. When he was five, Henry's mom bought him a little brown puppy with floppy ears. One day Henry and his friends decided to play catch with the puppy by tossing it back and forth over a picket fence. The dog banged into the pickets over and over. The pup died a couple of days later. Henry told me that just thinking about it now makes him want to cry. When I asked Linda if she had participated in animal cruelty as a child, she got very quiet and suddenly serious. She said yes but that she just could not talk about it. Ian was the least of the offenders. All he did was fry ants with a magnifying glass.

I was surprised that so many people I knew admitted to abusing animals when they were little. I should not have been. Animal rights law professor Justin Marceau points out in his book, *Beyond Cages: Animal Law and Punishment*, that a study found most of the people in a population of social workers and animal welfare workers admitted they had abused animals in the past. But none of my friends turned to the dark side—no felons, wife-beaters, or serial killers among them. Indeed, even Charles Darwin had a history of childhood animal abuse. He wrote in his autobiography that, as a boy "I beat a puppy, I believe simply from enjoying the power." (However, Darwin added, "This act lay heavily on my conscience as shown by my remembering the exact spot where the crime was committed.")

The idea that there is a strong link between childhood animal cruelty and violence directed toward people is so well established that the term "The Link" is now a registered trademark owned by the American Humane Association. Public presentations by Link advocates often begin with tales of tragedy. First, the school shootings: Parkland, Florida; Columbine, Colorado; Springfield, Oregon; Jonesboro, Arkansas; Pearl, Mississippi; Paducah, Kentucky—all committed by boys reputed to have abused animals. Then the serial killers: Albert DeSalvo (the Boston Strangler), Jeffrey Dahmer, Lee Boyd Malvo (the D.C. sniper accomplice)—all of them accused of childhood animal cruelty.

I have never been overly impressed with this type of anecdotal evidence. Some Link advocates would have you believe that most or even all serial killers and school shooters had a history of childhood animal abuse. Not true. An analysis of 354 cases of serial murders found that nearly 80% of the perpetrators did not have a known history of cruelty to animals. The connection between school shootings and animal cruelty is equally tenuous. Arnold Arluke and Eric Madfis analyzed rates of animal abuse among twenty-three young males involved in school massacres that occurred between 1988 and 2012. Most of the shooters had no history of animal cruelty. Indeed nearly 20% of them, had a record of pronounced empathy and affection for animals. The Sandy Hook Elementary School killer Adam Lanza, for instance, became a vegetarian because he did not want to harm animals. And Charles Andrew Williams, who killed two classmates and wounded thirteen others, apparently became extremely upset when one of his friends killed a frog. Williams kept field mice as pets and even had little funerals for them when they died.

Link advocates sometimes overstate the relationship between childhood animal abuse and adult mayhem. But there is evidence that the two are somehow related. Many studies have now compared rates of animal abuse in people with histories of violence and people with no history of violence. Dr. Emily Patterson-Kane used a statistical technique called meta-analysis to combine the results of fifteen of these studies. She found that there was some association between childhood animal cruelty and later violence; 34% of the violent offenders had a history of animal abuse compared to 21% of non-violent samples. But she was more impressed that most people who commit violent crimes against humans do not have a history of violence directed at animals.

How close is the relationship between animal cruelty in kids and adult aggression? There are several reasons why you might think that childhood animal cruelty and later violence would be connected. I call the first one the bad seed hypothesis. Some children are liars, cheats, thieves, and bullies by the time they are in elementary school. Psychiatrists refer to this pattern of behavior as a conduct disorder. In the 1960s, it was thought that three traits were especially characteristic of these kids: fire-setting, bed-wetting, and animal cruelty. While this trio is not as closely connected as originally thought, the American Psychiatric Association still includes "has been physically cruel to animals" as one of the diagnostic criterion for conduct disorders. The bad seed hypothesis holds that animal abuse is not a *cause* of delinquency. Rather, it argues that animal cruelty can be a sign of a troubled child, some of whom will become adult psychopaths.

A stronger version of Link thinking is called the violence graduation hypothesis. This is the idea that pulling the wings off butterflies and tying tin cans to a dog's tail are the first in a series of steps that can eventually end in prison. The title of an influential book by Linda Merz-Perez and Kathleen Heide captures this idea—*Animal Cruelty: Pathway to Violence Against People*. One implication of this theory is that animal abuse in children can be used as a form of profiling, a red flag that can identify potential serial killers and school shooters before their violence escalates. For example, PETA staffer Jennifer O'Connor wrote in the *Washington*

Post, "The FBI has identified cruelty to animals as a sign of psychopathy, a red flag indicating a high risk for committing acts of violence that pose a danger to the entire community."

The problem is that there is little evidence that animal abuse is an effective predictor of later violence. Researchers from Pacific University examined risk factors associated with involvement in school shootings. By far, the biggest risk factor was a fascination with guns, followed closely by being bullied, high interests in violent music and media, social isolation, having antisocial peers, and depression. Animal cruelty was, by far, the factor least associated with committing a school massacre. Similarly, Suzanne Goodney Lea, a sociologist, studied the backgrounds of 570 young adults. Fifteen percent of them had a history of animal abuse. The children in her sample who habitually lied, got into fights, used weapons, or set fires did tend to become violent adults. But this was not true of the kids who had abused animals. She concluded, "Animal cruelty, when considered against a range of other antisocial childhood/teenage behaviors, showed no significant predictive power in discerning which individual will go on to engage in human-directed violence as an adult."

There are other reasons we should be careful about making causal connections between childhood cruelty and adult violence. In Philosophy 101 (Logic), you learn that all A's are B's does not imply that all B's are A's. Thus the fact that most heroin addicts first smoked marijuana does not suggest that most first-time marijuana smokers will become junkies. Similarly, even if every school shooter and every serial killer had abused animals as a child (a false claim), we could not logically conclude that every child who pulls the wings off a moth will become a murderer. (After all, there is every reason to suspect that all school shooters and all serial killers drank milk when they were little.)

Arnold Arluke has a talent—listening to people. He puts them at ease, and they tell him things they would not normally reveal. He would have been a good homicide detective. Arluke used this ability to delve into the minds of college students who had tortured animals. They weren't hard to find. He just asked students in his classes. The students he interviewed had poisoned fish with bleach, ripped the legs off flies, burned grasshoppers with lighter fluid, played Frisbee with live frogs. This statement by a woman he interviewed is typical: "It was like we didn't have anything to do and we were bored, so it's like, 'OK, Let's go torture some cats!'"

Arluke came up with a radical suggestion. He argued that for many children, animal cruelty is a normal part of growing up. He called it "dirty play." It's forbidden fruit, like cussing or smoking cigarettes. He thinks animal abuse enables children to play adult power games in secret. It also helps form bonds with your co-conspirators, your partners in crime. Granted, the types of childhood cruelty that Arluke uncovered in his presumably normal sociology students were generally not of the microwaving cats and dropping puppies off the roof variety. And, unlike the hard-core criminals interviewed by Felthous and Kellert, most of Arluke's students felt remorse for their youthful indiscretions. But the fact remains that childhood animal abuse is more common than is generally recognized. One study found that 67% of male college students and 44% of female students admitted in an anonymous survey that they had abused animals.

The awkward fact is that most wanton animal cruelty is not perpetrated by inherently bad kids but by normal children who will eventually grow up to be good citizens—architects, guidance counselors, college professors. For me, animal abuse raises a big question, but it is not why deranged psychopaths are cruel. The answer to that question is obvious; they are mentally ill, morally blind, or evil. No, a more important issue transcends our relationships with animals: Why do fundamentally good people do fundamentally bad things?

For some researchers and animal protection organizations, the connection between animal cruelty and human violence is a moral crusade to be pursued with missionary zeal. A growing number of researchers, however, have come to question simplistic Link thinking. They worry that Link advocates and the media are perpetuating an irrational moral panic among the public. Link skeptics don't argue that we should ignore animal abuse. Rather, they believe that we should treat animal abuse as a serious problem in its own right, not because it turns children into adult psychopaths.

Anthrozoologists are divided over the strength of the connection between animal abuse and human violence. This debate is no different than contentious issues in other areas of human behavior. For years, psychologists have argued over whether violent TV causes aggression in children, whether pornography fuels sex crimes, and whether day care is good or bad for children. Like these questions, the controversy over the causes and effects of animal cruelty is not going away. The issue is too important.

■ ■ ■

As you can see from these examples, the scope of anthrozoology is broad. We investigate issues like personality differences between cat and dog people simply because they are interesting. On the other hand, studies of the effectiveness of animal-assisted therapy, and the link between animal cruelty and adult aggression have important real-life implications. But there is another reason that the study of our interactions with other species is both fascinating and important. It is that the ways we treat animals offer an unusual window into human nature.

As I show in the next chapter, the French anthropologist Claude Lévi-Strauss got it right when he wrote, "Animals are good to think with."

The Importance of Being Cute

WHY WE THINK WHAT WE THINK ABOUT CREATURES THAT DON'T THINK LIKE US

It's easier to empathize with the dog than with the flea.
—ERIC GREENE, PSYCHOLOGIST

Judy Barrett of Greensboro, North Carolina, had a problem. She and her husband were crazy about bluebirds. They spent a lot of money to entice a pair of them to nest in their backyard, even purchasing a snake-proof bluebird nest box and special bird baths. Judy even kept a supply of mealworms in her refrigerator because she heard bluebirds love them. The Barretts' amenities did attract a pair of birds, just not the kind they wanted. When they weren't looking, a common sparrow hijacked the nest box and laid five eggs in the bluebird house. Not knowing what to do, Judy sent a letter to Randy Cohen, who wrote the "The Ethicist," a Dear Abby-style advice column on everyday moral questions in the *New York Times Magazine.*

Would it be ethical, Judy asked, for her to destroy the eggs of the lowly sparrow to make room for a pair of lovely bluebirds?

Cohen said no. "In ethics, cuteness doesn't count."

Logically, of course, he is right. But while cuteness does not count in the rarefied world of philosophers, it matters a lot in how most people think about the treatment of other species. For instance, one of the biggest factors in how much money people say they would donate to help an endangered species is the size of the animal's eyes. This is bad news for the rare giant Chinese salamander. It is the largest and possibly most repulsive-looking amphibian on earth, a beady-eyed six-foot-long mass of brown slime. You don't see pictures of them gracing the fund-raising brochures of environmental organizations. But contrast the giant salamander with another Chinese animal, the equally rare but infinitely more appealing giant panda, whose eyes are framed by giant dark circles. Their gaze is so endearing that the panda is the logo of the World Wildlife Fund.

Of the 70,000 species of mammals, birds, fish, reptiles, and amphibians on the planet, only a handful merit much human concern. Why do we care about the giant panda but not the giant salamander, the eagle but not the vulture, the bluebird but not the common sparrow, the jaguar but not the Dayak fruit bat (one of two species of mammals in which the breasts of males produce milk)? The ways that we think about animals are often determined by species characteristics—how attractive or smart the creatures are, their size, the shape of their head, whether they are furry (good) or slimy (bad), and how closely they resemble humans. Too many legs or not enough legs are negatives. So are disgusting habits like eating feces or sucking blood. How an animal's flesh tastes also counts, though not as much as you might think.

The inconsistencies that haunt our relationships with animals also result from the quirks of human cognition. We like to think of ourselves as the rational species. But forty years of research in cognitive psychology and, more recently, behavioral economics shows that our thinking and behavior are often completely illogical. For example, in the Nobel Prize-winning psychologist Daniel Kahneman's book *Thinking, Fast and Slow*, he describes a study in which people were asked how much they would give to prevent waterfowl from being killed in polluted oil ponds. On average, they said they would pay $80 to save 2,000 birds and $78 to save 20,000 birds.

What is it about human psychology that makes it so difficult for us to think consistently about animals? The paradoxes that plague our interactions with other species exist because human thinking reflects a convoluted mire of instinct, learning, language, culture, intuition, and mental shortcuts.

BIOPHILIA: ARE HUMANS NATURAL ANIMAL LOVERS?

In an elegant little book written in 1984, the evolutionary biologist E. O. Wilson hypothesized that humans have a built-in attraction toward the natural world. He called this trait "biophilia." Some of our relationships with other species do seem to be innate. When I give talks about the psychology of human-animal relationships, I usually include a couple of slides that inevitably evoke a chorus of *oohs* and *ahs* from the audience. The pictures are of adorable kittens and puppies. The audiences' responses to the photographs reflect a component of human nature that makes behavioral scientists squirm: instinct. The notion that humans are innately drawn to anything that looks like a baby—infants, puppies, ducklings, you name it—is called the "cute response." The idea was first proposed by the Austrian ethologist Konrad Lorenz. Young animals have features in common with human infants: large foreheads and craniums, big eyes, bulging cheeks, and soft contours. Lorenz referred to these characteristics as "baby releasers" because they automatically bring out our parental urges.

Bambi is a classic example of how easily we are manipulated by baby releasers. Walt Disney originally urged the animators working on the film to draw the fawn as accurately as possible. He had a pair of fawns shipped in from Maine and made his artists watch an anatomist dissect the rotting carcass of a newborn deer. The problem was that the Bambi sketches that the animators produced, while realistic, were not cute enough to grab the hearts of the movie-going public. The solution was babyfication; Disney told the artists to reduce the length of Bambi's muzzle and make Bambi's head bigger. Then they gave Bambi huge eyes with lots of white in them. Bambi was morphed into a surrogate human baby.

Mickey Mouse is a similar testament to Disney's ability to design characters that elicit our parental urges. Mickey started life in 1928 as a not-so-nice trickster named Steamboat Willie. Over the next fifty years, Disney systematically changed Mickey's image. To accomplish this shift to a kinder and gentler Mickey, his features became more baby-like. Mickey's head grew to nearly half the size of his body, and the size of his eyes and cranium nearly doubled. Does our innate tendency to be taken in by a pair of oversized eyes affect our attitudes toward the treatment of other species? Of course. Stephen Jay Gould, the late Harvard biologist who traced Mickey's evolution, said it best: "We are, in short, fooled by an evolved response to our own babies and we transfer our reaction to the same set of features in other animals."

The role of cuteness in our attitudes toward the treatment of animals is illustrated by public outrage over the annual "harvest" of baby harp seals on the ice floes off the Atlantic coast of Canada. The seals are irresistible right after they are born; for the first two weeks of their lives, their fur is pure white and their eyes dark and as deep as Arctic pools. In the 1970s and 1980s, gory photographs showing the oozing blood of newborns being clubbed to death were staples of the brochures and placards of anti-hunt protesters. Eventually, the Canadian government caved into public pressure—sort of. They prohibited killing seal pups under fourteen days old, which happens to be when their fur becomes darker and the animals begin to look less infantile. After that, it is open season. The Canadians did not stop the baby seal hunt; they stopped the *cute* baby seal hunt.

Our fetish for animals that look like infants comes at a cost. Humans' love for the cute has produced canine breeds in which full-grown dogs are more like perpetual puppies. The babyish snouts of breeds like Chinese pugs and French bulldogs make for respiratory problems, and their bulging puppy eyes barely fit into their shallow sockets. By breeding dogs for neoteny (the biological term for the retention of juvenile features in adults), we have also created pets that are emotionally immature and prone to canine versions of our neuroses. This phenomenon has been a boon to Big Pharma, which has developed repackaged versions of Zoloft and Paxil for our depressed, anxious, and obsessive-compulsive pets.

WHY DO PEOPLE HATE SNAKES?

But if people are *biophilic* toward creatures like puppies and baby seals, they are *biophobic* toward others—snakes, for instance. In a Gallup poll, Americans were asked about the things that make them sweat. Four of their top ten fears were of animals, with snakes at the top of the list. (The other common animal fears were of spiders, mice, and dogs.) Even the revered medical missionary Albert Schweitzer, whose philosophy emphasized reverence for all life, kept a gun around to shoot snakes.

When I was a grad student studying animal behavior, I observed the conflict between the fascination for snakes and the fear of them up close and personal. I was spending the summer recording the grunts, hisses, and bellows of alligators and crocodiles at a reptile theme park in Florida. At the beginning of the tourist season, the park would hire college students who had experience handling reptiles to lead tours of the facility. At the end of each tour, the guide would don a pair of snake-proof boots and wade into a pit containing a dozen big rattlesnakes and water moccasins, snakes that can kill you.

The balloon trick was the climax of the show. The tour guide would blow up a balloon and pick out a big diamondback, which he would harass with a snake hook until the animal was riled up, coiled, and ready to strike. To pull off the trick, you have to hold one end of the balloon in your hand and slowly push the other end toward the hot snake. Then, when the balloon is a foot from the snake's head, you make a quick thrust, pushing the balloon directly into its face. If you do it right, the snake hits the balloon full-force with its fangs. Bang! The startled tourists jump and clap, and maybe even give you a big tip.

But one of the college boys did not have the guts to make the quick final thrust toward the inch-and-a-half fangs of an Eastern diamondback.

The old-timers on the staff did not think much of the summer college kids, especially this one. In the mornings, before the place officially opened, I would join them around the pit to watch the new guy try to learn the balloon trick. Looking confident and cocky in his starched khaki jungle shirt, he would enter the pit, pin a snake, grab its head, and milk it by hooking the snake's fangs over a glass vial and massaging its venom glands. No problem. But then it was time for the finale—the balloon trick. You could see his hands start to tremble when he started to blow up the balloon, and the tremors would get worse as he picked out his target.

That's when the old guys would start in on him, some of them clucking like a chicken, a few whispering encouraging words, "Come on, kid . . . you can do this." Then the college boy would line up the balloon and start to push it slowly toward the snake. But slow doesn't do it for a rattlesnake. You've got to be quick to make them strike. You've got to startle them.

The college kid would cautiously edge the balloon closer and closer to the snake's face until it touched the animal's nose, pushing the snake backward out of its strike coil, making a rattlesnake packing enough venom to kill five men look about as tough as a pussycat. Not a good way to impress the audience.

The kid, humiliated, would leave the pit, eyes down, the old-timers clucking, showing no mercy. On day seven of snake tour-guide training, the kid did not show up for work, and I never saw him again. The incident reminded me of the Apostle Paul's warning that the spirit might be willing, but the flesh is weak. On those morning sessions at the reptile park, primordial fears of the flesh prevailed.

Venomous snakes were a serious threat to our ancestors, and in some parts of the world, they remain a hazard to life and limb. About 100,000 people, primarily in Asia and Africa, die each year from snake bites, and three times that many suffer amputations or permanent disabilities. Objectively, however, fear of snakes among Americans does not make sense. The chances of dying from a snake bite in the United States is about one in fifty million; and most victims are testosterone-fueled males with more balls than brains. A case in point was described in an article in the *Annals of Emergency Medicine*. A forty-one-year-old man showed up at a hospital emergency room with a rattlesnake bite on the end of his tongue. The report speaks for itself. "A friend held the snake close to the patient's face while the patient mimicked the tongue protrusions of the reptile. Seizing the opportunity, the rattlesnake quickly bit him on the dorsal surface of the tongue. While the fangs were still in place, the friend yanked the snake out of the patient's mouth." Ouch. The man's tongue swelled to the size of an orange, making it nearly impossible for him to breathe, and he almost died.

Why are so many Americans afraid of snakes? After all, only about six people in the United States die annually from a snake bite, compared to thirty-four people killed by dogs. Is ophidiophobia a relic of Bronze Age mythology featuring serpents, a naked woman, and apples? Or are people weirded out by the snake's alien leglessness or its phallic form? Or did snake phobias evolve because they steered our ancestors away from animals that could kill them?

Scientists have been arguing for two hundred years about the relative importance of nature and nurture in the development of snake fears. The psychologist Susan Mineka argues that in monkeys, fear of snakes is learned. In her studies, rhesus monkeys captured in the wild were terrified of snakes but monkeys born in captivity showed no fear of them. However, if lab-reared monkeys observe one of their wild-caught brethren freak-out in response to a snake, they immediately catch a permanent case of snake-phobia.

Other investigators, however, disagree that nonhuman primates are blank slates when it comes to snakes. At the Kyoto Primate Institute, the ethologist Gordon Burghardt tested captive Japanese monkeys in a situation in which they had to reach in front of a snake cage to get food. Many of the animals were terrified of the snakes though they had never seen one before. In her book *The Fruit, the Tree and the Serpent: Why We See So Well*, the primatologist Lynne Isbell makes a convincing case that the visual system of nonhuman primates was shaped by evolution to specialize in snake detection.

In a series of studies, Vanessa LoBue and her colleagues tested the idea that humans have a built-in snake detector. They asked adults to pick out photographs of snakes embedded in a series of pictures of other natural objects. Sure enough, their subjects were quicker at spotting a picture of a snake amid pictures of other animals than they were to pick out a picture of a flower or a centipede. Then they repeated the research using three- and four-year-old children and even six-month-old infants. As with adults, even the youngest children were more attuned to snakes than other animals. LoBue believes that humans are not born with a fear of snakes but rather, what she calls "a perceptual bias for snakiness."

So nature plays a role in snake fears. But that can't be the whole story. About half of Americans say they are not afraid of snakes, and some 400,000 people in the United States keep them as pets. Further, cultures differ widely in how they treat snakes. My friend Bill spent five years in Tanzania as a game warden. In the village where he lived, people did not distinguish between poisonous and harmless snakes. When anyone saw a snake, they would shout "Nyoka!" and everyone would come running and help club it to death. But this is not true in New Guinea. According to the biologist Jared Diamond, New Guineans are not afraid of snakes despite the fact that a third of snake species on the island are highly venomous. Unlike Tanzanians, New Guinea tribesmen are adept at telling poisonous from nonpoisonous species, and they eat the harmless ones.

The idea that both genes and environment influence our attitudes toward animals fits nicely with E. O. Wilson's revised view of biophilia. While he originally conceived of biophilia as a hard-wired human instinctive urge to affiliate with all things bright and beautiful, he subsequently changed his mind. Wilson's new perspective emphasized the profound effects learning and culture have on our relationships with nature. "Biophilia," he wrote, "is not a single instinct but a complex set of learning rules that can be teased out and analyzed individually." And the job of teasing out the learning rules that govern our relationship with nature falls within the province of anthrozoology.

LANGUAGE AND MORAL DISTANCING:
WHAT'S IN A NAME?

How we think about animals is also affected by the names we give them and the words we use to describe them. Animal words permeate human language. Some of them elevate ("busy as a bee," "foxy lady!"), others demean ("you bitch"), and some reflect sexual power (*cock, pussy*). Calling someone "an animal" reflects our ambivalence about our place in nature. In some contexts, it is a compliment; in others, an insult. Psycholinguists argue about whether language reflects our perceptions of reality or helps create it. I am in the latter camp. Take the names we give the animals we eat. The Patagonian toothfish is a prehistoric-looking creature with teeth like needles and bulging yellowish eyes which live in deep waters off the coast of South America. It did not catch on with sophisticated foodies until an enterprising Los Angeles importer renamed it the considerably more palatable "Chilean sea bass."

The words we use for meat helps us avoid thinking about the ethical implications of our diet. It is easier to order a pound of beef from the butcher than a pound of cow. Semantic moral distancing is apparently less necessary as we descend the phylogenetic scale; we don't bother with linguistic cover-ups for chicken, duck, or fish. In other parts of the world, however, people dispense with meat euphemisms altogether. The German words for pork, beef, and veal are, respectively, *Schweinefleisch* (pig flesh), *Rindfleisch* (cow flesh), and *Kalbfleisch* (calf flesh). In Mandarin, beef is *niurou*, which translates into cow (*niu*) meat (*rou*); pork is *zhurou* (pig meat), and mutton is *yangrou* (sheep meat).

In recent years, the question of names for meat has become contentious and economically significant. At issue is the status of consumable flesh produced from stem cells taken from a living and unharmed animal. The meat is produced in Petri dishes and steel vats rather than from slaughtered animals. The first lab-grown hamburger was eaten in 2013 at a press conference in London. It costs an estimated $280,000 to produce the burger, which apparently did not taste very good. Since then, well-financed start-ups from Silicon Valley to Hong Kong have been racing to make these products both delicious and economically viable, but what should it be labeled? Big agricultural interests are in a tizzy. They call it "false meat," and they violently oppose any plant-based or animal-cell-based burgers, steaks, or seafood being labeled "meat."

The alt-meat industry is torn about what to call their product. A team of researchers studied how labels affected consumer attitudes toward flesh derived from stem cells. The term "clean meat" was the winner. It was perceived as healthier, tastier, and more natural than "animal-free meat," "cultured meat," or "lab-grown meat." In contrast, the loser, "lab-grown meat," was viewed as unnatural, dirty, and disgusting. And consumers also said they would be more willing to try a clean meat burger than they would a lab-grown Big Mac.

Partisans on both sides of the animal rights debate realize the power of words. In describing the Canadian seal hunt, the government agency that oversees the hunt uses neutral words—"harvest," "cull," "management plan." The language of seal hunt opponents is peppered with hot words: "slaughter," "massacre," "atrocity." What the wildlife managers call "the swimming reflex of dead animals," the activists refer to as "being skinned alive."

The animal rights group People for the Ethical Treatment of Animals (PETA) has made millions of Americans aware of the suffering associated with factory farming, hunting, animal research, zoos, and circuses. But they have had little success in riling the public up over the suffering caused by our insatiable desire for sushi-grade bluefin tuna or the pain experienced by a sixteen-inch brown trout who mistakes a #14 Royal Wulff dry fly for a real insect. My friend Cathy says she never eats anything with a face, but she doesn't count fish. PETA's strategy to change the way we think about creatures with fins rather than fur is to rename them. The slogan for their new campaign against fishing: "Save the Sea Kittens!"

Joan Dunayer, author of *Animal Equality: Language and Liberation*, agrees. She correctly believes that words make it easier for us to exploit other species. She proposes linguistic substitutions such as "aqua-prisons" for aquariums, "inmates" for zoo animals, and "cow abusers" for cowboys. She wants us to refer to our pets as "my dog friend" or "my cat friend." I am happy to call Tilly "my cat friend," but I suspect that my dentist who has an aquarium in his waiting room will be reluctant to say that it is time to change the water in his "fishy friends' aqua-prison."

PETS OR RESEARCH SUBJECTS? CATEGORIES COUNT

The language we use to talk about animals is closely tied to another factor that affects how we think about animals—the categories we put them in. For example, animals in the category "pet" are named; animals in the category "research subject" are usually not. When I asked a biologist if any of the mice in his lab had names, he looked at me as if I were crazy. I wasn't surprised. After all, the white mice he pokes, probes, and injects are essentially identical. Why should they merit names?

But sometimes our animal categories become blurred. When I was a graduate student, we did give names to some of our animal research subjects—the lifers, the ones that had become more pets than objects to be used in experiments. Our favorite lab animal was a five-foot black rat snake named IM (pronounced *em*). We got him when he was just a baby. IM was unusual in that he had two heads and one penis (most snakes have one head and two penises). One head was named Instinct and the other one Mind. You can see why we gave him a nickname.

But the shift from lab experimental subject to a pet can come at a cost. A laboratory animal veterinarian told me about the time she "instantly fell in love" with a beagle puppy that was scheduled to be part of an experiment that would end in the animal's death. She quietly took one of the lab technicians aside and told him to swap animals, so another dog was euthanized in place of the beagle. She realized that the beagle had lived only because a person of authority had taken a shine to it. Even after several years she felt guilty about arbitrarily sentencing the other dog to a death sentence.

The human propensity for categorizing animals starts young. Researchers at Yale University showed pictures of unfamiliar animals like saigas and pangolins and objects such as luzaks (a gizmo that draws circles) and garfloms (a device to flatten towels) to preschool children and recorded the types of questions they asked about them. The children's questions reflected a deep-seated category system that distinguished between living creatures and inanimate objects. When shown a pangolin, they would ask questions like "What does it eat?" and when presented with a garflom, they asked "How does it work?" or "What is it for?"—questions they never asked about the animals.

Is the human mind wired to think about animals differently than inanimate objects? In her book, *Naming Nature: The Clash Between Instinct and Science*, Carol Kaesuk Yoon describes a series of fascinating cases of people with brain damage whose mental capacities are intact except that they can no longer recognize and name animals. J.B.R., whose brain was damaged when he contracted encephalitis, could easily identify inanimate objects like flashlights, wallets, and canoes, but was completely stumped if you showed him a picture of a parrot or a dog. Researchers have also reported that some parts of your brain light up when you see pictures of animals but not pictures of human faces or inanimate objects. These studies suggest that parts of the human brain may have evolved to specialize in processing information about animals. Whether this degree of brain wiring is due to nature or nurture is unknown.

WHERE BUGS ARE PETS AND DOGS ARE PESTS: THE SOCIOZOOLOGIC SCALE

There are big differences between the way zoologists classify animals and the categories the rest of us put them in. While the phylogenetic scale is based on an organism's evolutionary history, in everyday life we look at animals in terms of what Arnold Arluke calls the sociozoologic scale. This is a sometimes arbitrary category system based on the roles animals play in our lives. For example, while dogs and rats lie on the same large branch of the phylogenetic scale, they are worlds apart on the sociozoologic scale. And the sociozoological scale has moral consequences. A team of social psychologists asked a sample of four hundred Americans to indicate whether they had moral concerns about the treatment of twenty species—*food animals* such as pigs and chickens, *appealing wild animals* such as chimpanzees and kangaroos, *unappealing wild animals* such as bats and snakes, and *pets* including cats and dogs. Nearly all the participants were concerned about pets, and 70% cared about appealing wild animals. In contrast, only half of them indicated they had any concern at all about the treatment of food animals, and even fewer cared about unappealing wild species.

Culture plays a major role in how we construct our sociozoologic scales. Take insects. Americans typically view invertebrates with a combination of fear, antipathy, and aversion. In Japan, attitudes toward creepy crawlies are more complex. Not many American children would jump for joy to receive a stag beetle for their birthday. In Japan, a lot of them would. The Japanese have a word, *mushi*, that is hard for Westerners to completely understand. For older Japanese, *mushi* refers to insects, spiders, salamanders, and even some snakes. To them, tadpoles are *mushi*, but adult frogs are not. Younger Japanese restrict *mushi* to insects, particularly singing crickets, fireflies, dragonflies, and big beetles with massive horns.

Mushi are male things. Boys catch them, keep them in elaborate cages, and even stage *mushi* strength contests. Tokyo department stores sell *mushi* collecting gear, *mushi* breeding material, *mushi* terrariums, *mushi* mattresses, and, of course, the bugs themselves, which can cost hundreds of dollars. Popular *mushi* activities include staging matches to see whose beetle can pull the most weight and provoking beetles to fight over pieces of watermelon—an insect version of sumo wrestling. You can watch these battles on YouTube. The Japanese word for a dog or cat is *petto*. Is a rhinoceros beetle *petto* or a toy? Eric Laurent, an anthropologist who has studied *mushi*, argues that in some important ways, these insects are pets—children play with and get obvious pleasure from their bugs.

Many children refer to their giant beetles as *petto*. One culture's pest can be another culture's pet.

The anthrozoologist James Serpell has developed a simple and elegant way to look at cultural differences in how we think about different species. He argues that our attitudes toward animals boil down to two dimensions. The first is how we emotionally feel about the species ("affect"). On the positive side, there is love and sympathy, and on the negative side, there is fear and loathing. The other dimension is "utility"—whether the species is useful or beneficial to human interests (perhaps we eat it or use it for transportation) or detrimental to our interests (for example, it eats us or the tomatoes in our garden).

Imagine a grid with four quadrants. The emotional dimension is represented by a vertical line with love/affection on the top and loathing/fear on the bottom. It is bisected with a horizontal line representing the utility dimension—the left side is "not useful/detrimental to our interests," and the right side is "useful." The grid now forms a four-cell category system that helps us think about the roles of animals in our lives and the categories we put them in: loved and useful (upper right); loved and not useful (upper left), loathed and useful (lower right), loathed and detrimental (lower left).

This simple category system even applies to cultural differences in attitudes about man's best friend, the dog. Guide dogs for the blind and therapy dogs fit into the "loved and useful" category. The typical American pet dog, on the other hand, is loved but is not particularly useful, at least in the traditional sense. In Saudi Arabia, dogs are generally despised; they exemplify the "loathed and detrimental" category. Perhaps the most interesting category consists of animals that are both loathed and useful. Dogs living with the Bambuti people of the Ituri Forest are derided, beaten, kicked around mercilessly, and left to scrounge for offal. However, the same dogs are considered valuable assets, as the Bambuti would be unable to hunt without them.

Serpell's model also offers a perspective on shifts in our attitudes toward a species. In an article titled "How Pigeons Became Rats," Colin

Jerolmack examined the depiction of pigeons in *New York Times* stories over 150 years. He found that, in the minds of New Yorkers, pigeons have shifted from the "liked but not useful" category to the "loathed and not useful" category. This change also describes how my brother-in-law feels about deer. When he first moved into his home on a bluff overlooking the Puget Sound, he loved seeing the deer stroll through his backyard. They reminded him of Bambi. Everything changed when, to the delight of the hungry deer, he put in a vegetable garden. Now he hates them, and Bambi has joined rats and geese (they poop on his lawn) in the sociozoological quadrant that contains the loathed and useless.

IN ANIMAL ETHICS, HEART TRUMPS HEAD

How we think about animals also reflects a perennial theme in human psychology—the conflict between logic and emotion.

On the afternoon of September 3, 1977, a twelve-foot-long Nile crocodile named Cookie was spending Labor Day weekend doing what crocodiles do best: basking on its belly in the sun. Cookie lived at the Miami Serpentarium, a reptile theme park that was home to hundred-year-old tortoises, pythons big enough to swallow goats, and an array of exotic lizards and venomous snakes. Among the many visitors to the park that late summer day were six-year-old David Mark Wasson and his father. Eager to catch a glimpse of the crocodile, they edged close to Cookie's pen and saw him lying still by the pond in his cage. Mr. Wasson decided to show his son that crocodiles do move. He set David on top of the pen's concrete wall and looked around for a couple of wild grapes to throw at Cookie. You probably can guess what happened next.

The instant Wasson turned his back, David fell into the pen, directly on the spot where Cookie was usually fed. Large crocs can move like lightning when they want to, and it took about a millisecond for Cookie to grab the little boy. When Bill Haast, the park's owner, heard the screams of the crowd, he ran toward the pen, vaulted over the wall, and immediately began pounding on Cookie's head with both fists. Tragically, he failed to wrest David from the 1,800-pound reptile, and Cookie slithered back into his pond with David in his jaws. The boy's body was recovered several hours later.

Haast was devastated. Late that night, he climbed into the crocodile's pen and pumped nine bullets from a Luger into Cookie's head. It took him an hour to die.

When I heard about the deaths of David and Cookie, the logical part of me thought that the execution made no sense. Though he weighed nearly a ton, Cookie's brain was the size of my thumb. It is safe to say that a crocodile is not what philosophers refer to as a "moral agent." After her husband shot Cookie, Haast's wife said, "The crocodile was just doing what comes naturally to him." She was right.

Still, another part of me, a more primitive part, understood the need for retribution. So did the *New York Times* editorial writer who described the croc's death as "emotionally satisfying yet thoroughly irrational." Was shooting Cookie the right decision? In this situation, should we listen to logic, which says that it is not appropriate to punish a crocodile for acting on its instincts, or to our emotions, which cry out for revenge for the death of an innocent child?

The debate over whether human morality is based on emotion or reason goes back a long time. The eigtheenth-century philosopher David Hume argued that morality was the product of our emotions, while Immanuel Kant believed human ethics were based on reason. When I first became interested in the psychology of human-animal relationships, I decided to find out what goes through people's minds when they think about moral issues involving other species. At the time, the field of moral psychology was dominated by the Harvard psychologist Lawrence Kohlberg.

Like Kant, Kohlberg believed that moral decision making was based on thoughtful deliberation: We weigh the pros and cons of a course of action and then we make a logical decision. Kohlberg's research focused on the development of moral thinking in children. He would tell them a story that included a moral dilemma. Then the children would make a judgment about the situation and explain their reasons. The classic Kohlberg scenario was the case of Heinz, a poor man who steals an overpriced drug from a greedy pharmacist to save his wife, who is dying of cancer. In deciding whether Heinz was right to steal the drug, Kohlberg's kids were little logicians. They considered factors like the chances Heinz would get caught or the happiness that would accrue from his wife's survival.

Shelley Galvin and I used a variation of Kohlberg's method to investigate how people make decisions about the use of animals in research. Our study was simple. The participants evaluated a series of hypothetical proposals for experiments involving animals. We asked them to approve or disapprove each experiment and to explain the reasons for their decision. In one case, a researcher pursuing a treatment for Alzheimer's disease wanted approval to implant stem cells taken from monkey embryos into the brains of adult monkeys. In another, a scientist sought permission to amputate the forelimbs of newborn mice to study the roles that genes and experience play in the development of complex movement patterns. Both scenarios were based on real experiments.

About half the participants approved the monkey study, while only a quarter of them thought the mouse amputation study was OK. We were not surprised by their decisions, but we did not anticipate their reasoning. In the case of the monkey experiment, the students' thinking tended to be rational. They based their decisions on considerations such as the costs and benefits of the research or the intrinsic rights of the animals. Not so when it came to cutting the legs off mice. In this case, the participants wrote statements like, "This experiment repulses me." "Think of the expression on the poor little animal's face!" and "Gut-wrenching!" Our subjects based their judgments about amputating the forelimbs of baby mice not on logic but on their emotional reactions to the experiment.

Based on the prevailing theory of moral development in psychology, we had mistakenly assumed that our subjects would take the rational route in making their decisions. Instead, we found they often listened to their guts. This finding would have been predicted by Jonathan Haidt, one of the leaders of a school of moral psychology that emphasizes the primacy of heart over head in ethics. Like many psychologists, Haidt believes that human cognition involves two distinct processes. The first is intuitive, instantaneous, unconscious, effortless, and emotional. The second process, in contrast, is deliberative, conscious, logical, and slow. It kicks in only after we have made our quick intuitive decision and cleans up the cognitive mess by coming up with justifications for our emotion-based decisions.

Haidt argues that our moral judgments reflect the operation of these two systems, and that in matters of morality the nonlogical intuitive system usually predominates. Haidt's theory of morality was nicely captured by Lucy, a special educator I interviewed for a study of animal rights activists. When I asked her about the importance of logic and emotion in her path to animal protectionism, Lucy said, "It always stems from the emotional. But a lot of times I have to find an intellectual rationalization for my emotional reactions. Otherwise, I can't sway people or defend my position."

MORALITY, ANIMALS, AND THE YUCK FACTOR

Like Lucy, we can usually come up with some sort of justification for our moral judgments. But sometimes logic flat-out fails. Haidt asked people to consider a series of situations that were highly offensive yet caused no harm. In one, a woman cleans a toilet with an American flag. In another, an adult brother and sister on vacation in Europe decide to have sex one time using two forms of birth control. One of his scenarios involved a pet: "A family's dog was killed by a car in front of their home. They had heard that dogmeat was delicious, so they cut up the dog's body and cooked and ate it for dinner."

You make the call. Is it OK for them to throw the family dog's corpse on the grill?

When people are asked if it is permissible for the family to eat their pet, most of them immediately say, "No! It is not OK to eat your own dog!"

The problem comes when you push them on their reasoning—when you ask them to explain exactly why it is wrong to eat an animal that is already dead and incapable of feeling any pain. Most of the time, they simply can't come up with any reasonable justification for their decision. It's the Yuck Factor. The act just seems disgusting.

The University of Pennsylvania psychologist Paul Rozin calls disgust the moral emotion. Disgust elicitors such as having sex with a sibling are universal across human cultures. Bodily products such as feces, urine, and menstrual blood are repulsive to people, no matter where they are raised. According to Rozin, they are disgusting because they remind us of our animal origins. But nurture factors such as social class can also affect our moral intuitions. Haidt found that 80% of poor Philadelphians said that people should be stopped from eating their dead dog, but only 10% of upper-class educated Philadelphians felt that way. He believes wealthier individuals tend to operate under moral systems that emphasize whether an act causes harm as opposed to its offensiveness—and in this case, the dog was already dead.

There is, of course, a difference between what people think and what they do. I suspect even his wealthiest subjects would never actually order a Philly cheesesteak sandwich made with onions, Cheez whiz, and chopped beagle meat.

SHOULD WE ALWAYS SAVE PEOPLE OVER ANIMALS?

The first death caused by a self-driving car occurred on the morning of March 18, 2018. Elaine Herzberg was walking down a street in Tempe, Arizona, when an Uber SUV operating on "fully autonomous mode" slammed into her at forty miles an hour. Uber suspended their self-driving car trials, but autonomous vehicles were soon back on the streets in cities like Miami, San Francisco, Pittsburgh, Toronto, and Washington, D.C. Self-driving cars are essentially intelligent computers on wheels programmed to make instantaneous decisions. Sooner or later, some of these decisions will fall into the realm of ethics. Should, for example, cars like the one that killed Elaine Herzberg be programmed to swerve away from a child if it means running over an old person . . . or a dog?

Seeking to discover universal ethical principles that could be programmed into self-driving cars, investigators from MIT developed an online computer game called the Moral Machine Experiment. Published in the journal *Nature*, their results shed light on the relative value humans around the globe place on pets as opposed to people. Here's how the game worked. On a computer screen, players viewed a series of self-driving cars with broken brakes. Their job was to choose the better of two courses of action for the car. For example, a player might have to decide whether a self-driving car should swerve into the left lane and kill a child crossing the street or continue in the right lane and smash into two elderly men. Some of the decisions involved sparing the life of a dog, a cat, or a person.

After translating their game into ten languages, the researchers put it online so anyone with a computer could play. To their astonishment, the game went viral and generated forty million decisions from people in two hundred countries. Three "essential building blocks of machine ethics" emerged from their massive data set. These were the principles agreed on in nearly every culture. The most important of these was "Spare people over animals." (The other two were "Spare more lives over fewer lives," and "Spare the young over the old.")

If you are thinking "Aha! The self-driving car game is based on the classic trolley problem ethicists use to study ethical thinking," you would be right. Indeed, in a series of studies, the psychologist Lewis Petrinovich used the trolley problem to find out how our moral decisions differ when human interests are directly pitted against those of other species. Here are the original versions. What would you do?

Version 1. A trolley's brakes have failed, and it is speeding down a set of tracks toward five people. You can save them if you pull a switch that will divert the trolley onto a different set of tracks where one person is standing. That person will be killed if you pull the switch. Is it morally permissible to divert the trolley and prevent five deaths at the cost of one human life?

Version 2. The runaway trolley is headed for five people. This time you are crossing a footbridge over the tracks. Right next to you is a large man. You can save the five people if you push the man off the bridge into the path of the trolley. Is this morally permissible?

If you are like most people, you made different decisions in these two cases. In Version 1, 90% of people say yes—you should throw the switch and divert the trolley so one person dies rather than five. But in Version 2, only 10% feel that shoving the large guy into the path of the trolley is the right course of action.

Why do people usually make different decisions in these cases? After all, the outcome is exactly the same: One person will die and five will live. I posed these two trolley problems to one of the most moral persons I know, my wife. Mary Jean made the same decision that most people do. But when I probed her reasoning, it came down to intuition. She said that throwing a switch to save someone feels a lot different from pushing a person off a bridge. Why? Using brain imaging technology, the neuroscientist Joshua Greene found that the personal version of the trolley scenario (pushing the person) causes the emotional processing centers of the brain to light up while the impersonal version (throwing the switch) does not.

Petrinovich also used trolley problems to find out how our moral decisions play out when human interests are directly pitted against those of other species. Here are two of Petrinovich's animal trolley scenarios:

Version 3. An out-of-control trolley is headed toward a group of the world's last five remaining mountain gorillas. You can throw a switch and send it toward a twenty-five-year-old man. Should you?

Version 4. The trolley is speeding toward a man whom you do not know. But you can throw a switch and send it hurtling toward your pet dog? Should you?

In both cases, Mary Jean said to save the person over the animal, even if it would mean the death of Tsali, our late, great Labrador retriever. I made the same decision, and you probably did too. Petrinovich found that almost everyone chooses to save people over animals in these situations. This is also true of people in different parts of the world. In fact, of all the ethical principles he examined using many different types of trolley problems, Petrinovich found that the single most powerful moral rule was "Save people over animals."

There are, however, exceptions. Researchers at Georgia Regents University asked people if they would save a dog or a person if both of them were in the path of a runaway bus. The vast majority said they would save, for example, a foreign tourist rather than a dog. But if they were told the dog was *their personal pet*, 30% of men and 45% of women indicated they would let the bus run over the foreign stranger. The zoologist Harry Greene feels the same way. He told me he once forked out $4,000 in emergency veterinary bills for Riley, his yellow Lab, whom he described as the kind of dog you only get once in a lifetime. Harry handed a Visa card to the vet and said, "Save the dog." He said he did not feel a shred of remorse for saving Riley rather than spending the money to feed starving children in Darfur.

MENTAL SHORTCUTS AND ANIMAL ETHICS

Laurie Santos, director of the Yale University Canine Cognition Center, calls human thinking "glitchy" because it often relies on quick and dirty rules of thumb.

Quick—Answer these questions.

1. A bat and ball, in total, cost $1.10. The bat cost $1 more than the ball. How much does the ball cost?

2. Are you more likely to die from a shark attack or being hit by a part falling from an airplane?

If you are like me, you said ten cents for the first question and shark attack for the second one. The correct answers, however, are five cents and the airplane part. The reason that you were probably wrong is that our thinking often relies on quick and dirty rules of thumb that cognitive psychologists call heuristics. Heuristics are efficient and they usually produce correct solutions to problems. I use heuristics on Sunday mornings when I go up against the *New York Times* crossword puzzle, and doctors use them when they decide whether an emergency room patient is suffering from a heart attack or indigestion. But these mental shortcuts can bias our thinking and lead us astray.

Human moral thinking relies on similar rules of thumb. Some of these moral heuristics have their roots in evolution: for example, our aversion to incest or to betraying friends. Our predilection for senseless revenge is the result of the inappropriate application of what the legal scholar Cass Sunstein calls the "punishment heuristic." This principle explains the irrational belief that killing Cookie, the boy-eating crocodile, was justified.

One of the most important heuristics is called framing. It refers to the fact that the way we think about a problem is affected by how it is posed. Mental frames are affected by cultural norms and our cognitive habits, and they determine how we view situations. Once we have a problem framed, we don't consider alternative explanations or solutions. Framing helps explain one of the most troubling of all the paradoxes of human-animal relationships, the Nazi animal protection movement.

HOW THE NAZIS COULD LOVE
DOGS AND HATE JEWS

"He was too!"

"He was not!"

"Yes, he was!!"

"No, he wasn't!!"

My colleague Laura Wright and I were shouting at each other in front of a class of students at Western Carolina University. We were guest speakers in a course on film and literature, and our disagreement was over whether Adolf Hitler was a vegetarian. The students had watched the documentary *Food, Inc.*, and their teacher had invited us to discuss our perspectives on meat with the class. Laura is a highly regarded scholar in a new academic field called vegan studies. We are good friends, but we have different perspectives on the consumption of animals. She does not eat them. I do.

The class was fun, and the students seemed engaged. Laura showed a video clip of Lisa Simpson's conversion to vegetarianism, and I read part of an essay I had written on the horrors of industrial chicken production. Things were going fine until I used the Nazi animal protection movement to illustrate how cultures can twist moral values.

I inadvertently pushed Laura's buttons when I casually remarked that Hitler was a vegetarian and that he once told a female companion who was eating a sausage, "I didn't think you wanted to devour a dead corpse, the flesh of dead animals." Laura was not buying it. She said Hitler regularly devoured liver dumplings and stuffed squab. That's when we started yelling at each other.

I suspect Laura is right. But a bizarre moral inversion did occur in prewar Germany that enabled reasonable people to be more concerned with the suffering of lobsters in Berlin restaurants than genocide. In 1933, the German government enacted the world's most comprehensive animal protection legislation. Among other things, the law forbade any unnecessary harm to animals, banned the inhumane treatment of animals in the production of movies, and outlawed the use of dogs in hunting. It banned docking the tails and ears of dogs without anesthesia, the force-feeding of fowl, and the inhumane killing of farm animals. The law also restricted animal experimentation. Hitler signed the legislation on November 24, 1933. This was only the first in a series of Nazi animal protection acts. In 1936, for example, the German government dictated that fish had to be anesthetized before slaughter and that lobsters in restaurants had to be killed swiftly.

In announcing restrictions on animal research in a 1933 radio address, Hermann Göring said, "To the Germans, animals are not merely creatures in the organic sense, but creatures who lead their own lives and who are endowed with perceptive facilities, who feel pain and experience joy and prove to be faithful and attached." Göring once threatened, "I will commit to concentration camps those who think that they can continue to treat animals as property."

Hitler objected to killing animals for scientific research and he believed that hunting and horse racing were "the last remnants of a feudal society." As you might expect, contemporary animal activists don't relish the idea that Adolf Hitler was a fellow traveler, and, like my vegan pal Laura, some animal activists adamantly deny that he was either a vegetarian or an animal lover. But the anthrozoologist Boria Sax has carefully documented the evidence that many leading Nazis, including Hitler, were genuinely concerned about the treatment of animals. (Needless to say, the fact that Hitler loved animals does not in any way undermine the validity of the case for animal protection.)

The Nazis used framing to construct a perversely inverted moral scale in which Aryans were at the top and Jews were classified as "subhumans"—beings lower than most animal species. While German shepherd dogs and wolves were high on the moral hierarchy, Jews were compared to lower animals—rats, parasites, bedbugs. In 1942, Jews were forbidden to keep pets. In one of history's great ironies, the Nazis followed the legal procedures governing humane slaughter when they euthanized thousands of Jewish pets. But, unlike their dogs and cats, Jews were not covered under German humane slaughter legislation. No, they were sent to concentration camps, where their treatment would not pass muster under the Third Reich's animal welfare laws. For the Nazis, Jews blurred the boundaries between man and animal. They were a polluted class, freaks who were neither fully human nor completely animal.

To me, Nazi animal protectionism is the ultimate paradox in our relationships with other species. It speaks volumes about human moral thinking. A few pages ago, I argued that for a thousand generations, the genetic puppet-masters have murmured into our ears "people over animals." Hitler's ability to construct a culture in which dogs were afforded moral status denied to Jews and homosexuals illustrates the fact that with enough social pressures, humans will ignore the whisperings of the genes. The problem is that the power to resist our genes does not necessarily make us better people.

ANTHROPOMORPHISM: WHAT WE THINK ABOUT WHAT ANIMALS THINK

Nazi animal protectionism exemplifies the twisted ways that humans sometimes think about the relative moral status of people and animals. But nutty thinking about other species can pop up anywhere. A couple of years ago, I was kayaking down the Nantahala River, a popular whitewater stream in western North Carolina. In the summer, it is jammed with rafts of paddle-flailing tourists trying to avoid rocks and cross-currents. The river is beautiful and frigid, forty-five degrees all year long. You don't want to fall in.

I had paddled halfway down the river when I caught a whiff of cigar smoke coming from a raft a hundred yards in front of me. There was an obese fifty-year-old man in the raft puffing the offending cigar and guiding his wife and a little brown Chihuahua through the rapids. The dog was not having a good day. She was shivering uncontrollably and looked terrified. And that was before their raft flipped over.

I have to give the guy credit. He kept the cigar clamped between his teeth even after he was dumped out of the raft. The little Chihuahua deserved credit too. She had the sense to climb onto the nearest large floating object—Cigar Man. That's how they went down the river, a man with a soggy cigar jammed in his mouth, his wife, and a hypothermic dog desperately clinging to the man's head. I remember wondering what made that guy think his Chihuahua would enjoy running the rapids of a freezing Class III river. The answer is anthropomorphism. Humans are natural anthropomorphizers. It is part of our mental equipment. In the 1940s, psychologists showed that humans will even attribute motivations to animated geometric figures moving around on a movie screen—"Now the red triangle is *really* pissed at the blue square. You go, girl!"

A stunning example of the human need to project our desires and emotions and mental states onto other creatures played out in 1999 when the Sony Corporation began marketing a series of interactive robotic dogs named AIBO, a compound of *artificial intelligence robot*. With its shiny metallic body, AIBO looked to me more like a friendly space alien than a puppy, but it walked like a dog, and it could cuddle and play and respond to sounds. AIBO would even let its owner know if it was happy or ticked off. At about $2,000 a pop, AIBO was not cheap.

Researchers from the University of Washington and Purdue University investigated how children and adults responded to an AIBO compared to a real dog. The researchers found that some individuals became very attached to their robotic puppies. One adult owner, for example, admitted to an online discussion group that he felt embarrassed getting dressed in front of his AIBO. Another wrote, "I love Spaz [his AIBO]. I tell him that all the time. . . . When I first bought him, I was fascinated by the technology. Since then, I care about him as a pal. I do view him as a companion. . . . I consider him to be a part of my family. He's not just a toy. He's more of a person to me."

The responses of children to AIBO also illustrate the connection between anthropomorphism and the attribution of moral status to animals. Most of the kids in these studies said that AIBO experienced emotions like happiness and could see a toy and hear what you are saying to it. They also felt it would be wrong to punish AIBO if it did something wrong, or to hit AIBO, or throw it away. When asked why, the children said things like "Because he will be hurt," and "Because he will cry until you come back." And all of the children said you should do something to help AIBO if he gets hurt or if his tail breaks off.

THE PROBLEM WITH HAVING A THEORY OF MIND

Our tendency to project ourselves into a robotic pet came hand in hand with the evolution of a big brain. The capacity to infer the perspectives of other people and to mentally put ourselves in their shoes would have been a huge advantage to our ancestors. Their success in the Darwinian race to pass on their genes hinged on the ability to forge political alliances, vie for mates, and figure out whom they could and could not trust. The ability to imagine what other people are thinking and feeling is referred to as having a "theory of mind." Humans have this ability, but whether even large-brained animals such as chimpanzees and dolphins might have it is debated.

When we anthropomorphize, we are extending our theory of mind to members of another species. This tendency lies at the heart of many of our moral quandaries with animals. Take hunting. James Serpell points out that the hunter who can think like a wild pig is more likely to come home with the proverbial bacon. But the hunter who sees the world from the point of view of an animal he is trying to kill would automatically empathize with it and, hence, feel guilty for killing it. My game warden friend Bill lived in an African village where baboons would destroy crops and create mayhem. The villagers would trap them in pits at night and kill them the next morning, but they felt bad about it because their eyes looked so human. They have a saying in Swahili, "Never look a baboon in the eye." It makes it too hard for you to kill them.

Do the metaphorical roots of original sin lie in two conflicting attributes of human nature: our inclination to empathize with animals and our desire to eat their flesh? Serpell eloquently lays out the moral issues our big-brained ancestors faced: "Highly anthropomorphic perceptions of animals provide hunting peoples with a framework of understanding, identifying with, and anticipating the behavior of their prey. . . . But they also generate moral conflict because if the animals are believed to be essentially the same as persons or kinsmen, then killing them constitutes murder and eating is the equivalent of cannibalism."

Anthropomorphism is the source of much of our guilt over the treatment of animals, but there is another problem with projecting our mentality onto other species. Our interpretation of their behavior is often wrong. The perpetual smiley faces of the dolphins at SeaWorld indicate that these animals are happy spending their lives swimming in endless circles around a concrete tank. Wrong. The yawn of an alpha male baboon means he is bored. Wrong. (He is using the display of his formidable canine teeth to say, "I can rip your face off.") When Tilly gently rubs her face on my leg, she is showing that she loves me. Wrong. She is scent-marking my legs with glands on her cheeks, telling the world that she owns me.

Researchers at the University of Portsmouth found that half of British dog owners say their pets feel shame and guilt. You know the look—tail between legs, the sorrowful eyes that won't look you in the face saying, "I didn't mean to poop on the rug." Dixie, our Golden retriever, would break your heart when she gave you what our veterinarian called the "tragic look." But does the guilty look, that hangdog expression, really signify that your dog knows he or she has sinned?

According to the canine scientist Alexandra Horowitz, the answer is probably not. As she describes in her book *Inside of a Dog*, Horowitz devised an ingenious experiment to tell whether dogs get the guilty look when they *actually* misbehave or when their owners *think* they misbehaved. In the experiment, owners instructed their dogs to not eat a dog biscuit that was placed right in front of them. Then the owners briefly left the room. In some cases, the experimenter then gave the treat to the dog; in others, she just took it away. In half the trials, the owners were told that their dog had disobeyed them when, in reality, the dog had done absolutely nothing wrong. (I know; it seems unjust.) The results indicated that dogs only gave the sad look when their owners *thought* they had disobeyed, not when the dogs ate the biscuit. The experiment does not prove that dogs lack a moral sense. It does, however, show that we can easily misinterpret their expressions and behavior.

Our tendency to impute mental states to other creatures depends on the species. As the psychologist Eric Greene noted, "It is easier to empathize with the dog than the flea." Shelley Galvin and I once devised a scale to measure the degree to which people attributed mental states such as consciousness and pain to species ranging from ants to chimpanzees. The seventeenth-century French philosopher René Descartes argued that nonhuman animals were mindless biological robots, but our subjects disagreed. Over 90% of them believed dolphins, chimpanzees, dogs, and cats experienced moderate to high levels of consciousness. But attributions of mental states dropped considerably once you got past pets and creatures with big brains. Only 60% of our subjects thought that pigs possessed at least a moderate degree of consciousness. Yet comparative psychologists have discovered that the cognitive capacities of pigs are comparable to dogs and even primates. Bats fared even worse. Despite the fact that they are among the most social of mammals, only 30% of people believed bats had much in the way of mental experiences.

People, of course, differ in what they believe goes on inside the heads of animals. Take a question that came up on a LISTSERV for fishermen—do fish feel pain? For an angler identified as L.C. the answer was clear: "Fish have no nerves in their mouth and cannot register pain from being hooked. They feel tension, like someone holding your arm gently on a public sidewalk." But another fisherman disagreed: "Let's put it this way," he wrote. "If I stuck a fishhook in your mouth and pulled you through the water, would it hurt? Does that answer the question?"

Anthropomorphism is a deeply rooted component of human nature, and anthropomorphic thinking has been tied to specific regions in the human brain. Importantly, beliefs about animal mental experiences have moral consequences. British researchers found that individual differences in a trait they called BAM (Belief in Animal Mind) are strongly related to attitudes toward the treatment of other species. Acknowledging that animals can feel pleasure and pain can cause some people to be more concerned about animal welfare than others. But the causal arrow can point in the other direction. Studies have shown, for example, that eating animals makes people more likely to deny the existence of mind in creatures such as cattle and pigs.

WHAT IS IT LIKE TO BE A SPIDER?

Ethologists—researchers who study animal behavior—are in a tough position when it comes to understanding animal minds. On the one hand, they come home to tail-wagging dogs and know their puppies are happy to see them. But for good reason, they are uncomfortable when it comes to speculating about the subjective worlds of the spiders, octopuses, snakes, and elephants that they study.

In a classic article titled "What Is It Like to Be a Bat?" the philosopher Thomas Nagel argued that we can never know what it is like to be a bat or, for that matter, any other animal. Not all animal behaviorists agree. I remember as a graduate student attending a session on primate behavior at the International Ethology Congress in Kyoto, Japan. At the end of the last presentation, one of the big guns in the field stood up and asked a question. "Before we leave," he said, "I would like to ask how many of you went into the field of animal behavior because you wanted to know what it is like to be a member of the species you study." I was at the back of the room thinking to myself, what a stupid question. I was completely wrong. More than half the scientists in the audience raised their hands.

Over the last twenty years, a flourishing field of cognitive ethology has developed, among whose intellectual tools are what the psychologist Gordon Burghardt calls critical anthropomorphism. These days, animal behaviorists talk of empathy in mice, negotiations in chimpanzees, and post-traumatic stress in elephants. I once asked Fred Coyle, an arachnologist, what he thought went on in the minds of the spiders he studied. For example, do they plan the architecture of their webs? Or are their muscles and glands mechanically following the dictates of genetically programmed neural impulses? I could tell that my question made Fred uncomfortable. "Hmm," he said. After a long pause, he said he thought of spiders as robots—predatory AIBOs with eight legs.

Fred went to graduate school at Harvard, and his officemate, also an arachnologist, did not feel the same way about the minds of spiders. He really did want to know what goes on in their heads. One afternoon, he borrowed a large playpen from a friend and bought yards and yards of stretchy rubber tubing from a hardware store. Then, weaving the rubber tubes around the playpen's frame, he carefully constructed a huge web modeled on the web-building techniques of the spiders that he studied.

Late one night, Fred unexpectedly returned to their lab to pick up a book he needed. There, in the dark, he found his friend, crouched silently in the middle of the giant web, figuring out what it was like to be a spider.

■　■　■

The bottom line is that there are several reasons human-animal interactions are often paradoxical and morally incoherent. Thousands of studies in cognitive psychology have demonstrated that human thinking about nearly everything is surprisingly irrational. And the waters get particularly muddy when we think about other species. Instincts seduce us into falling in love with big-eyed creatures with soft features. Genes and experience conspire to make it easy for us to learn to fear some animals but not others. Our culture tells us which species are loveable, hateable, and eatable. Then there are the conflicts between reason and emotion, our reliance on hunches and empathy, and our propensity to project our own thoughts and desires into the heads of others.

No wonder our relationships with other species are so messy.

3

Pet-o-Philia

WHY DO HUMANS (AND ONLY HUMANS) LOVE PETS?

> The human desire to associate with other animals, to observe them, and get inside their worlds, lies at the heart of pet-keeping.
>
> —JESSICA PIERCE, BIOETHICIST

> If we recognize the intrinsic value of animals' lives, then it is immoral to keep them for our pleasure, whether we call them companions or pets.
>
> —LESLIE IRVINE, SOCIOLOGIST

Antoine, a young Frenchman in his early twenties, approaches an attractive young woman in a pedestrian mall. With him is a cute dog named Gwendu, which in Brittany means "white and black."

"Hello," he says to her. "My name's Antoine. I just want to say that I think you're really pretty. I have to go to work this afternoon, but I was wondering if you would give me your phone number? I'll phone you later and we can have a drink together someplace."

She hesitates for a second, looks at him, says "*Oui*," and pulls a pen from her purse.

In truth, Antoine is not the man's name, Gwendu is not his dog, and he is not in the mall looking for a beautiful woman to hook up with. He is actually the confederate in an experiment designed by a pair of French anthrozoologists, Serge Ciccotti and Nicolas Guéguen, who is Gwendu's owner. They are studying the effectiveness of pets as social lubricants. Over several weeks, Antoine, who was selected for the experiment because a panel of women rated him as unusually attractive, approached 240 randomly selected young women. On half these approaches, he was alone and on the other half, he was accompanied by the dog, which researchers described as "kind, dynamic, and pleasant."

Did being associated with Gwendu increase Antoine's sexual chemistry? *Mais oui!* About 10% of the women gave him their phone numbers when he was by himself, but nearly 30% of them fell for Antoine's line when he was accompanied by *le chien*.

The anthropologist Peter Gray and his colleagues took this line of research to the next level. They examined how customers of the online dating site Match.com used pets to attract romantic partners. A cardinal tenet of evolutionary psychology is that women tend to allocate more resources to child-rearing, while men devote more time and energy to mating. Gray predicted that women should be particularly attuned to how their dates treated their pets. Men, they reasoned, should be more apt than women to use their pets to attract sexual partners. They also predicted that interactions with dogs would provide more salient cues than those with cats about parental skills of a potential long-term partner. That's because dogs generally require more care and attention. Finally, they tossed in a couple of questions for fun—for example, they asked the women in the study, "What is the hottest pet a guy can own?"

In collaboration with the chain pet store PetSmart, the researchers surveyed 1,200 singles on Match.com who owned a pet. Their predictions were on the mark. Men were four times more likely than women to use a pet to attract dates on the site. More women than men said they were especially attracted to partners who owned a rescue animal. Women were also more likely to admit they would not date someone who did not like pets. As predicted, more dog owners than cat owners talked about their pets on the site. Finally, 85% of the women said a dog was the hottest pet a guy could own compared to 18% who found cat owners hot. (None of the women thought men with pet rabbits were sexy.)

So, pets—at least dogs—can help you get dates. But the fact that pets can be social lubricants does not explain why people bring cats, birds, turtles, and even rats into their homes and treat them like family members.

From an evolutionary point of view, pets are a problem. Why should humans invest so much time, energy, and resources on creatures with whom we share no genes and who do no useful work? Most pet lovers, after all, are not handsome young men looking to increase their reproductive potential. Pet industry marketing executives will tell you that we bring animals into our lives because they make us happier, healthier, and more loved. I think it's more complicated than that. Consider these two very different pet scenarios.

NANCY AND CHARLIE: WHEN THE BOND WORKS

The Japanese bombed Pearl Harbor a couple of months after Nancy and Roy Watson were married. Roy immediately enlisted in the army. They trained him as a radio operator and sent him to fight the Japanese in Okinawa. In 1946, Roy came home, and he and Nancy started carving out their chunk of the American dream. Roy got a job in the parts department of the local Ford dealership. They soon had two sons whom Nancy stayed home to raise. In a couple of years, the couple had saved enough money to buy a brick rancher in West Asheville with a shady fenced-in backyard. There was a lot of activity in the Watson household, but it did not include animals. Roy wasn't partial to dogs or cats, and Nancy never considered herself an animal person either. Ten years after he retired from Ford, Roy died of cancer, leaving Nancy alone in the house where they had lived for forty years. Roy's death left a hole in Nancy's heart. A year later, she was depressed and felt that her life was empty. Her sons were worried about her. They didn't think she could last long rattling around alone in the empty house. There was talk of an assisted-living facility.

A year to the day of Roy's death, Nancy stopped at a convenience store to pick up a loaf of bread and noticed a handwritten sign at the checkout counter—KITTENS. The girl at the register asked Nancy if she wanted to see them. She said no and drove on home. She did, however, mention the kittens to her son, Aaron, who was visiting for the weekend. He said "Mom, let's take a look at them." To his surprise, she agreed. They found two seven-week-old kittens lying in the bottom of a cardboard box in the back of the store, one a calico, the other midnight black. She picked up the black one, and it was love at first sight. She named him Charlie.

Nancy and Charlie have been living together now for eight years. Nancy is spry, cheerful, and sharp. She tells me that she and Charlie are a team. "He has made my life happy," she says. "I've never felt alone since Charlie's been here. He is all I have." In addition to being her buddy, Charlie brings structure to Nancy's life. As soon as she wakes up in the morning, she makes their breakfast—a teaspoon of canned tuna for him, a bowl of cereal for her. Then he goes outside for ten or fifteen minutes. When he returns, he jumps on Nancy's lap, and they sit and talk for a while before Charlie moseys into the den for a nap. He wakes up in the afternoon, and, if it is sunny, they sit next to each other in matching chairs in the front yard until dark. She makes their dinner, and they eat together. Charlie does not like TV so he goes outside for the evening, but he comes back and checks on her two or three times before she goes to bed.

Nancy admits that there are disadvantages to having a cat for your best friend. While she is sleeping, Charlie morphs from Dr. Jekyll to Mr. Hyde. Then he takes to the woods and does what cats do at night: He hunts. He proudly brings Nancy the results of these forays: a freshly killed bird, lots of voles, a squirrel, and just last week, a baby rabbit. Sometimes the animals are mangled but breathing, and she opens the front door and tells Charlie to take them back outside. And he does.

Nancy and Charlie illustrate the best of the human-animal bond. Nancy's relationship with a cat has made her life immeasurably better, and I suspect that Nancy's good health, mental acuity, and ability to live at home by herself are due in part to their relationship.

"It has just been the two of us for eight years," she tells me.

The Nancy and Charlie story is not unusual. It is played out in millions of American homes every day. I saw it in my own parents who were never animal lovers until Pop retired, and they got the first of three dachshunds, all of whom they adored and all of whom they named Willie.

But, as Sarah Coe will tell you, living with pets is not always sweetness and light.

SARAH'S DOGS: WHEN THE BOND FAILS

When I interviewed Sarah, she was an administrator at a West Coast veterinary hospital and her husband, Ian, worked in information technology at a university. They had been married three years when Sarah decided they needed a dog. Ian was not enthusiastic, but eventually he came around. Though she had never had a dog herself, Sarah had encountered lots of animals with behavior problems in the course of her work at the veterinary clinic, so she and Ian went about choosing their puppy methodically. They spent months researching the characteristics of dozens of breeds before settling on a Shiba Inu, a foxy Japanese dog bred for hunting small game. The National Shiba Club of America uses terms like "spirited boldness," "very lively," "macho stud muffin," and "fiery little fuzzballs from hell" when describing the breed's temperament. For Sarah and Ian, this turned out to mean trouble from the start.

Hiro was nine weeks old when they selected him out of a litter in Oregon. He was high-maintenance from the get-go. If they didn't give him constant attention, he would shriek inconsolably, and he became unmanageable unless they exercised him for an hour and a half a day. Fortunately, there was an off-leash dog park near their house, but Hiro was socially inept and could not figure out how to act around the other dogs. His nerdiness soon created conflicts with the other dog owners.

One afternoon, Hiro, then six months old, decided to play dry-hump with a young Tibetian terrier, also a male. Ian knew that play mounting is a common behavior in young dogs and that it indicates nothing about their sexual orientation. The terrier's owner, however, started screaming, "No one humps my dog!! No one humps my dog!!"

To no avail, Ian calmly tried to explain to the terrier's owner that the puppies were just playing. But the conversation quickly escalated into a shouting match.

After a series of similarly unpleasant incidents, Sarah and Ian grew tired of other dog owners lecturing them about their pet's bad behavior. They stopped taking Hiro to the park and began paying a professional dog walker $300 a month to take the dog for long runs so they could get an hour or two of peace at home.

One of the veterinarians Sarah worked with suggested that Hiro's ADHD might be alleviated if he had a playmate. Big mistake. Nami, their new Shiba puppy, was even crazier than Hiro. She was a bully—unpredictable, aggressive, and demanding. Nami was so jealous that Ian and Sarah had to sneak into bed at night. She would even get upset when Ian kissed Sarah each morning when he left for work. By the time she was two years old, Nami was on both Valium and Prozac. The drugs helped some, but she was still a nut case. Most dog owners think of their pets as their children. Ian and Sarah's children were a surly punk and a borderline psychotic.

Sarah liked a tidy house, clean floors, and furniture that matches. All that changed with the dogs. They chewed up the sofa, destroyed the rugs, and generally wreaked havoc. "I don't want our house to look like a crazy person lives there," Sarah told me. Ian and Sarah are pleasant and fun to be around, but the dogs ruined their social life. The couple quit having friends over for dinner because Hiro and Nami would bark incessantly and try to steal food from the table.

Despite the fact that the dogs were ruining their lives, Sarah and Ian were genuinely attached to them. Sarah made outfits for Nami. Ian identified with Hiro. Ian told me that he and Hiro were both round pegs trying to fit into a square world. The Coes tried obedience school and consulted with some of the country's best dog behaviorists. Nothing worked. Several times a year, they would discuss getting rid of the dogs—adopting them out, even euthanasia. But their timing was out of sync. When Sarah was ready to throw in the towel, Ian was too attached. A few months later, their perspectives would reverse.

I asked them if the dogs where taking a toll on their relationship, and there was a long pause. They looked at each other and said, yes, that they had started seeing a marriage counselor. A week after our interview, Sarah sent me an email. She and Ian had decided to separate. Sarah was temporarily moving into an apartment. She said that the stress of living with the dogs from hell was a big factor in the breakup of their marriage. It is unclear what will happen to Nami and Hiro.

FAMILY MEMBERS OR PROPERTY?

As Nancy and Sarah show, sometimes the human-pet relationship works out and sometimes it does not. But what exactly is a pet? The historian Keith Thomas argues that pets are animals that are allowed in the house, given a name, and never eaten. This is a good place to start, but there are exceptions. My neighbors never let their dog inside, and my dentist has not named his tropical fish. Even the eating part has occasional exceptions. One evening when I was a graduate student studying alligator behavior in Florida, Mary Jean and I dropped in on our friend Jim, a retired agriculture professor whose property bordered the lake that was the base of my operations. Jim kept a menagerie around his mini-farm—goats, bantam chickens, Muscovy ducks, a couple of peacocks, a Chinese pug, and some guinea pigs that were his kids' pets. His wife and kids were away for the weekend and Jim was making himself dinner. As we chatted, he nonchalantly took a guinea pig from its cage, bopped it on the head with a stick, skinned it, and put it on the grill. I guess he didn't consider it a pet.

I prefer James Serpell's definition of pets. He says pets are animals we live with that have no obvious function. But even with this loose definition, things get weird around the edges. Until relatively recently, most animals in American homes had some sort of job. Dogs, for instance, were expected to herd, hunt, guard, pull carts, and even churn butter. Cats were tolerated more as biological mouse traps than objects of affection. In the United States, animals whose only function was to amuse their owners were rare until the mid-nineteenth century when there was an explosion in the popularity of caged birds, primarily singing canaries.

The range of animals that humans have kept as pets is extraordinary—crickets, tigers, pigs, cows, bears, rats, alligators, giant eels—the list is endless. But when asked what animals they consider pets, most people don't say eels or crickets. What do they think of? The answer, of course, is dogs and cats.

Cognitive psychologists refer to an item that exemplifies a category as a prototype. Right now, think of a bird . . . I suspect you conjured up the image of a sparrow, a robin or an eagle rather than an emu or a penguin. That's because robins are more "birdy" than ostriches. Which animals are the prototypes of pets? Samantha Strazanac and I asked college students to rate sixteen types of animals—all of which are sometimes kept as pets—on how "pet-ty" they were. Everyone, of course, said that dogs and cats exemplified the concept of "pet." Goldfish came in third; 75% of the students felt they had a high degree of "pet-ness." Only about half the students thought of hamsters, gerbils, rabbits, parrots, and parakeets as pets. Mice and iguanas scored low on our pet-ness scale, and white rats and tarantulas were even further down the list. Boa constrictors came in dead last; only 5% of the respondents considered them pets.

Some animal advocates don't like the word "pet." They find it demeaning to the animals we live with. They want us to call our furry, finned, and winged friends *companion animals* and their owners *guardians*. But substituting the term *guardian* for *pet owner* is problematic. Unlike the guardian of a human child, a pet's "guardian" is allowed to give away, sell, or castrate their wards against their will. Legally, they can even have their companion animal euthanized if they tire of it. The terms companion animal and pet guardian are linguistic illusions that enable us to pretend we do not own the animals we live with.

As the bioethicist Jessica Pierce argues in her book *Run, Spot, Run: The Ethics of Keeping Pets*, bringing animals in our homes because they give us pleasure poses a host of ethical issues. In developed nations, most dogs and cats are not allowed to roam free which deprives them of their right to liberty. Further, we routinely neuter them which eliminates their opportunity to enjoy the pleasures of sex and parenthood. Indeed, the animal rights legal scholar Gary Francione rejects the institution of pet-keeping altogether because it rests on the assumption that animals are property.

If pets are property, we should be able to put a value on their life. There are several ways to measure the relative monetary value of, say, a pet dog and a cat. Colleen Kirk, professor of marketing at the New York Institute of Technology, asked dog and cat owners to imagine that their pet had contracted a serious illness. Her subjects were then told, "It turns out that curing your pet requires much surgery, and without the surgery, your pet will die. How much would you spend for the life-saving surgery for your pet?" On *average*, her subjects said $10,700 for their dog and $5,200 for their cat. However, those numbers were thrown off by a handful of people who said $100,000. In this case, a more accurate measure is the median—the midpoint. Again, dog owners in the study were willing to pay almost twice as much to save their pets ($3,000) than were cat owners ($1,800).

These numbers, however, don't take into account the monetary value of the emotional toll of the death of a pet. For that, we can turn to the legal system. The state of Tennessee, for example, has established that, for the purposes of the law, a dog is worth $3,000. However, some juries see things differently. In 2014, police officer Rodney Price was investigating a burglary in Glen Burnie, Maryland, when he was confronted by a Chesapeake Bay retriever named Vernon owned by Michael Reeves. Feeling threatened by the big dog, Price pulled out his gun and shot him. While officer Price was exonerated by an internal police investigation, Reeves filed a wrongful death lawsuit. The jury agreed and awarded him $1.26 million in damages for Vernon's death.

The University of Colorado sociologist Leslie Irvine, author of the book *If You Tame Me: Understanding Our Connection with Animals*, is torn over the moral status of pets. She writes, "If we recognize the intrinsic value of animals' lives, then it is immoral to keep them for our pleasure, whether we call them companions or pets." At an intellectual level, she believes it is immoral to breed and imprison animals for our personal pleasure, and she argues that, because they are legally a form of property, the human-pet relationship is more like slavery than true friendship. The problem is that Leslie is deeply attached to her dogs and cats. Thus, she goes on, "I dread the thought of coming home to an empty house, no tails wagging in excitement to see me. . . . But my pleasure in being greeted and kept company does not justify keeping a supply of animals for that purpose." Leslie is confronted with the classic conflict between head and heart, and as is usually the case, heart wins.

TURNING PETS INTO PEOPLE

Like Leslie, a lot of Americans are in love with their animals. According to the American Pet Product Manufacturers Association, 63% of American households include a pet. In 2019, we shared our lives with ninety million dogs, ninety-five million cats, twenty million birds, nine million reptiles, and one hundred sixty million fish. Kasey Grier, author of *Pets in America: A History*, has found that the popularity of household animals in the United States goes through periodic surges. One occurred in the late nineteenth century when pets became a symbol of domestic tranquility and were viewed, particularly by mothers, as a way to instill responsibility and an ethic of kindness in children. The post-World War II years saw another uptick of interest in bringing animals into American homes, this time fueled by the growth of the suburbs and the idea that having a pet was a necessary component of a normal childhood.

Americans have been living with pets on and off for a couple of centuries, but beginning twenty or thirty years ago, the human-pet relationship seems to have entered a new phase—what the pet products industry refers to as the humanization of pets. Eighty-five percent of pet owners now tell pollsters that they consider their pets to be full-fledged family members. Seventy percent of pet lovers say they sometimes allow their animals to sleep in their beds, two-thirds buy their pets Christmas presents, 23% cook special meals for them, and 18% dress their animals up on special occasions.

Surprisingly, while our attitudes toward pets have changed, the proportion of American households that include a companion animal has not. Every five years or so, the American Veterinary Association conducts a national survey of pet ownership in the United States. In 1987, they reported that 38% of American homes included a dog and 31% included a cat. Their 2018 survey reported that dog ownership had remained 38% though cat ownership had dropped to 25%.

In recent years, however, the amount of money we dole out for the animals in our lives has skyrocketed. The American Pet Products Association estimates that we spend $100 billion a year on our furry and feathered friends. We spend more on our pets than we do on movies, video games, and music combined. This includes $40 billion for food and treats, $30 billion for veterinary care, $20 billion on live animals and pet supplies like kitty litter, designer dog clothes, collars and leashes, food bowls, toys, and birthday cards. We also fork out $11 billion for pet sitters, boarding kennels, washing and grooming services, obedience training, massage therapy, dog walkers, funeral urns, insurance policies, and New Age animal communicators.

The real action in the pet trade is at the high end. Luxury brands comprise 20% of pet food sales, but they generate over half of the industry's profits. These include menu items such as Fromm Nutritionals' Shredded Duck Entree ("generous portions of hand-shredded free-range duck simmered in its own natural duck broth with potatoes, peas, and carrots"). Your pet can wash down her gourmet dinner with Bowser Beer for dogs or enjoy Pet Refresh Bottled Water. The latest trend is for foods that are "all natural" and "organic." For example, Dr. Harvey's Homemade Biscotti for

Dogs is an all-organic treat made from oat flour, organic barley, honey, bee pollen, apples, dandelion, broccoli, peppermint, and fennel.

A lot of pet owners think their animals enjoy being dressed up. Tea-Cups, Puppies & Boutique, an Internet pet boutique, offers a fetching "Scarlet Cashmere Couture Dog Dress" for $920 as well as a lower-priced line of T-shirts, tank tops, jackets, and denim overalls. If you want to tote your pup around all day, TeaCups offers a pink "Wonder Nest Dog Carrier" for $1,649. A Manhattan-based pet products store appropriately named Bitch New York offered the "Liz Dog Collar Custom" for $990. For bikers who take their pooch to the annual "hog" rally in Sturgis, South Dakota, Harley-Davidson makes a line of dog motorcycle gear.

If your puppy needs some downtime, she might enjoy an afternoon at the L.A. Dogworks, a five-star dog resort that promises your pet a "total wellness experience." This includes an hour in the Zen Den, billed as a "simple Eastern retreat for your dog to relax and indulge." Many luxury hotels now offer services for pets. For example, your dog is welcome at the Sarasota Ritz-Carlton—if it weighs less than twenty pounds and you are willing to pay an upfront $250 pet fee. (In comparison, the pet fees in the Trump Hotel chain seem a bargain at $50 a night.)

The blurring of the boundary between pets and people is not a new phenomenon. The theme played out in nineteenth-century France as well. As the French middle class grew in size and influence, so did their fascination with pets, and within fifty years dogs and cats were transformed from working animals to family members. By 1890, luxury and pet ownership went hand-in-hand. The wardrobe closet of a well-decked-out dog in Paris might include boots, a dressing gown, a bathing suit, underwear, and a raincoat. Dog grooming salons sprung up in France as did pet cemeteries. At the beginning of the nineteenth century, dead dogs were dumped in the Seine, but by midcentury, Parisian animal lovers were having their deceased dogs and cats interred in pet cemeteries or their heads mounted for display over the mantle.

At least in the modern era, the term "humanization" applies more to dogs than the other species we bring into our lives. For example, Ben and

Jerry's is now marketing a line of dairy-free ice cream for dogs. In grocery stores, it will be placed next to the popsicles in the frozen food aisles— not in the pet food section. A Ben and Jerry's spokesman got it right when she said, "This is an opportunity for people to treat their dogs as they are treating themselves when they grab a pint off the shelf." This trend is not limited to the United States, as a woman named Alison Lever wrote to me in an email:

> Where I live in rural Spain, there has been a big change in attitudes to pet dogs over the last twenty years. Previously, they tended to be mutts from "accidental" matings and allowed to wander around the pueblo. Now they are more likely to be bought and kept in the home. The move to allowing dogs indoors seems to go with seeing them as part of the family.

Kasey Grier's book *Pets in America: A History* is the definitive history of the human-animal bond in the United States. I asked her what accounts for the extraordinary increase in the money we spend on our animals. She said that pets, and particularly dogs, are now framed by the pet industry as consumers themselves. An increasing number of people think their pets want and deserve the same stuff they have—biscotti, breath mints, raincoats, summer camp, all-natural foods, spa treatments. Even high-dollar weddings.

The question is who, aside from members of the American Pet Products Association, benefits from these excesses? Stephen Zawistowski, author of the book *Companion Animals in Society*, tells people, "If you buy a twenty-dollar coat for your dog to protect her from the cold, it's for the dog. If you buy her a $200 coat, it's for you." Morris Holbrook, professor of marketing at Columbia University, offers this advice to corporations trying to tap into the lucrative pet products marketplace: "Remind people that pets are not possessions; but rather animal companions with needs, wants, and rights comparable to those of other family members." Then he adds, "Assume that animal companions are basically people. You won't go far wrong."

Pets used to be a fairly inexpensive pleasure. Until the years after World War II, most household animals in the United States lived on table scraps and never saw the inside of a veterinary clinic. This is no longer the case. According to the American Kennel Club, the lifetime cost of owning a medium-size dog is $16,000. What kind of bang are you getting for your pet buck?

DO PETS PROVIDE US WITH UNCONDITIONAL LOVE?

I once surveyed pet owners at veterinary clinics about what they got out of their relationship with their pets. Three themes emerged loud and clear: "My pets are members of my family," "My pets are my children," and "My pets are my friends." In a follow-up study, social psychologist Robin Kowalski and I asked dog and cat owners to evaluate a series of statements comparing the benefits they derived from their relationships with their best human friend and the benefits derived from their relationships with their pets. Our subjects said that both human and animal friends were equally good at providing companionship, ameliorating loneliness, and making people feel needed. Human friends were better than animal companions when it came to being someone to confide in or talk to. There was one area in which pets clearly had the edge over human friends: providing unconditional love.

A glut of feel-good books tout the idea that pets shower their owners with unconditional love, but I believe this idea is overrated as an explanation of why so many humans live with animals. If pets provided unconditional love, you would think that everyone would be bonded to the animals in their homes. They are not. Timothy Johnson and his colleagues reported that 15% of adults in their study said they were not particularly attached to their pets. In informal polls I have taken in my classes, roughly a third of my students indicate that someone in their family fears, dislikes, or even hates the family pet.

The demography of pet-keeping also presents a problem for the unconditional love hypothesis. This theory of attachment to pets predicts that people living alone would be most in need of unconditional love and thus have high levels of pet ownership. This is not the case. According to a 2018 survey by the National Opinion Research Center, only 26% of American adults living alone had a pet compared to 53% of married couples. But, while adults with children have the highest rates of pet ownership, as a group, they are less attached to their animals than people who live alone. In fact, pet attachment drops a notch with each additional person in a family. Pets living in homes with young children really get the shaft. For example, about 25% of pets in families with children are groomed every day compared to nearly 80% of pets who reside with adults who do not have kids. Dogs and cats in childless homes are most likely to be showered with Christmas presents and go with the family on vacations. Sadly, the dog that was "our baby" during the first years of a couple's marriage is often demoted the moment their first child comes home from the hospital.

I have to admit that the unconditional love idea appealed to me more when we lived with dogs. But now that we are a cat family, I've had to reconsider. Tilly's love for me is entirely conditional. She calls the shots. She loves me when I make her dinner or give her a snort of catnip, and when she wants me to rub her belly or play a round of swat-Hal's-feet. But much of the time, I'm just the guy who opens the door when she wants to go outside.

DO PETS MAKE THEIR OWNERS HEALTHIER AND HAPPIER?

There are other reasons people might want to have animals in their lives besides the fact that they stroke our egos. Companion animals might be good for our health. Pet industry lobbyist Mark Cushing certainly thinks this is the case. In his book *Pet Nation: The Love Affair that Changed America*, he wrote, "There is no medicine more powerful, available to all and less expensive than having a pet—dog, cat, rabbit, bird, fish, hedge-hog . . . you name it." Over the last decade, pet industry marketing departments have relentlessly promoted that getting a pet will cure all your ills. Take, for example, the Human Animal Bond Research Institute. HABRI, as it is known in the industry, is a trade group whose stated aim is "gathering, funding, and sharing scientific research to demonstrate the positive health impacts of companion animals." HABRI wants to see a pet in every American home. The organization's president, Steve Feldman, told an interviewer, "Everybody should quit smoking, everybody should go to the gym, everybody should eat more fruits and vegetables, and everyone should own a pet. Those things all go into the same category, and we are going to put it there."

HABRI press releases tout the idea that overwhelming scientific evidence has demonstrated that pet ownership lowers blood pressure, reduces stress, prevents heart disease, decreases doctor visits, and makes people feel less depressed and lonely. An economic analysis commissioned by HABRI concluded that pet ownership saves Americans $12 billion a year in health-care costs. In addition to HABRI, the media also touts the idea that pet ownership is linked to improved mental and physical health. For example, a headline in *Time* magazine proclaimed, "Science Says That Pets Are Good for Your Mental Health," and books like *The Healing Power of Pets: Harnessing the Amazing Ability of Pets to Make People Happy and Healthy* become bestsellers.

The pet industry calls the idea that pet ownership comes with health benefits the "pet effect." Their marketing efforts have been spectacularly successful. According to surveys, over 70% of pet owners are now aware of scientific research showing that pet ownership can improve physical and mental health. A HABRI survey found that 97% of family practice doctors believe there are health benefits of living with a pet, and 60% of doctors interviewed said they had recommended getting a pet to some of their patients.

Unfortunately, there is a mismatch between the claims of corporate public relations machines and the actual results of empirical studies on the impact of pet ownership. As the authors of a recent review in the journal *Applied Developmental Science* wrote, "The mass media and the public seem to have an inexhaustible appetite for stories of animals helping people with their illnesses and disabilities. Unfortunately, satisfying this appetite often results in superficial and inaccurate media accounts of scientific findings." So, are pets four-legged miracle workers whose undying love can heal the human body and mind? Or are these claims mostly the result of industry hype and wishful thinking? Understanding this debate requires a bit of history.

The most important publication in the history of anthrozoology was a six-page article that appeared in the July 1980 issue of *Public Health Reports*. The lead author, Erika Friedmann, had just received her PhD in behavioral biology from the University of Pennsylvania. In her doctoral research, Friedman investigated the role of social support in survival from heart attacks. She asked ninety-six patients in a coronary care unit to complete a survey concerning their socioeconomic status, living situation, and their connections with their friends and family. She also threw in an item asking whether they lived with a pet.

Twelve months later, she tracked down the participants to see how they were doing. The big surprise was that owning a pet made a difference in their survival rates. While 28% of the non-pet owners had died by the end of the year, only 6% of the pet owners were dead. Excited by these results, Friedmann presented them at a meeting of the American Heart Association. The cardiologists, however, basically yawned, though one did refer to her study as "cute." The media was more interested. Her phone began to ring, and soon she was reading about herself in *Reader's Digest* and *Time*. It was a good way to start a career.

Erika's heart attack study inspired a flood of interest in the effects of the human-animal bond on human health and happiness that continues today. In the years following her study, some studies have, indeed, provided support for the pet effect idea. Researchers have found, for example, that children raised in homes with animals missed fewer days of school because of illness, and elderly people who live with pets have lower levels of cholesterol and depression. A study of over 10,000 Germans and Australians found that the participants who lived with pets made fewer doctor visits than non-pet owners. Researchers reported that Chinese women who owned dogs slept more soundly. A study at the University of Missouri found that a dog-walking program increased the level of physical activity of the participants. Some studies have shown that adults who are attached to animals are less lonely than their pet-less friends, and there is talk in some medical circles about prescribing dachshunds and Yorkies to old people.

NOW THE BAD NEWS

It would be nice to think that getting a dog or cat would cure all your ills, but don't throw out your Lipitor and Prozac just yet. One of the cornerstones of science is the notion of replication—the idea that other researchers can repeat your study and see if the results hold up. Scientists got a wake-up call in 2005 when John Ioannidis of Stanford University published an article with the provocative title, "Why Most Published Research Findings Are False." The paper prompted a rush to replicate established findings, particularly in psychology and clinical medicine. The results were not pretty. For instance, an international team of 200 researchers could not repeat the results of fourteen of the twenty-eight widely cited studies in psychology. And in 2012, the pharmaceutical company Amgen announced they could not replicate forty-seven of fifty-three landmark cancer papers. Ioannidis pointed out that some areas of science are prone to false findings. These include research areas in which studies tend to have low numbers of subjects and experimental treatments which have small effects. The bloom is also more likely to ultimately fade in areas of research that suddenly "get hot" as early results don't pan out as more and better studies are done.

Research on the connection between pet ownership and mental health is a good example. Take depression. Publicity materials put out by Zoetis Inc.'s "Pet Effect" publicity campaign claimed, "Research has shown that pets can help alleviate symptoms of depression. That's why therapy animals or pets are increasingly included in depression treatment plans." Internet health sites such as WebMD uncritically accept these claims by routinely posting articles with titles like "How Pets Help Manage Depression." But what researchers have found about the ability of pets to chase away the blues is way more complicated.

Keep in mind that you can't trust the findings of a single research project; you have to look for patterns of results across multiple studies. Between 1989 and 2020, thirty-three published studies compared depression levels in pet owners and non-pet owners. The patterns of the results were clear. Twenty-one of these studies found *no* differences in depression among people who lived with pets versus those who did not. Further, five of the studies reported that pet owners were *more* depressed than non-pet owners. Indeed, only five of the thirty-three studies found that pet owners were less depressed. These five studies tended to be of lower quality. For example, the studies in which pets made no differences in depression had, on average, ten times more subjects than studies which found pet owners were less depressed. In short, the preponderance of the evidence indicates that getting a pet has little, if any, impact on depression in their owners. Zoetis Inc.'s suggestion that dogs and cats be included in depression treatment plans is misleading.

Claims about the impact of pets on human physical health are equally problematic. A study of the impact of living with pets on the health of 42,044 randomly selected California adults is a good example. The study was conducted by a group of high-powered researchers and statisticians from the RAND Corporation. They analyzed data gathered as part of the ongoing California Health Interview Survey. About half of the participants lived with a pet, and they were equally divided between cat and dog owners. The researchers did find that both the dog and cat owners had somewhat better general health than people who did not have a pet. But there were also non-pet-related differences between people with and without pets. The pet owners were, for example, more likely to be younger, female, married, and white. They were also wealthier and tended to own their homes. When the researchers plugged these socioeconomic differences into their equations, the differences in health evaporated. The researchers concluded that health differences between pet owners and non-pet owners had nothing to do with living with dogs or cats, and everything to do with factors like age and money.

In a subsequent study, the RAND researchers examined the health of 5,000 children, half of whom had pets. As expected, the kids with pets were better off. They were healthier, more obedient, less moody—and they exercised and had fewer behavioral issues and learning problems at school, too. But the children with pets had lots of other advantages besides having a dog or a cat to play with; they lived in more stable families and were less likely to be on free school lunch programs or to have parents that did not speak English. More of them lived in houses with yards rather than in apartments, more of them were white rather than African-American or Hispanic. As with the adults, once these factors were taken into account, the differences between kids with and without pets disappeared.

THE IMPACT OF PETS ON HEALTH: CAUSE OR EFFECT?

The RAND Corporation studies illustrate *the causal arrow problem* in claims about the impact of pets on health and happiness. It's what statistics professors drum into the heads of their students—"Correlation does not imply causality." A slew of large epidemiological studies has found that pet owners (particularly, dog owners) have fewer health problems. But, as with the RAND research, these studies also show that pet owners are more likely to be younger, female, and more financially secure than non-pet owners. So, while it is possible that pet-keeping causes better health, it is equally possible that having better health, more energy, and more money causes people to welcome a dog or cat into their lives. For these reasons, a scientific report from the American Heart Association concluded there is a possible association between dog ownership and cardiovascular health. But they also cautioned, "Pet adoption, rescue, or purchase should not be done for the primary purpose of reducing cardiovascular disease risk."

"Pet Effect" studies are also prone to replication problems. The anthrozoologist John Bradshaw has pointed out that many of the early studies showing that pet owners were heathier than non-pet owners have not panned out. A University of Missouri dog-walking study found that while the exercise program was enjoyable for the participants, it did not lower their blood pressure, nor did they lose weight. A study of 21,000 people in Finland found that pet owners had higher blood pressure and cholesterol levels than non-pet owners. The pet owners in the Finnish study were also more susceptible to kidney disease, arthritis, sciatica, migraines, and panic attacks. And, while the Finnish pet owners did not smoke and drink as much as non-pet owners, they exercised less often and were more likely to be overweight. Researchers at the Australian National University found that adults between the ages of sixty and sixty-four who lived with a pet consumed more pain medications and were in worse mental and physical shape than people without pets. Another study compared elderly people who regularly played with pets with a control group who rarely or never played with animals. The death rates in the two groups were the same; and playing with pets made no difference in the participants' health and sense of well-being. (The people who played with pets did, however, drink more alcohol.)

Some results have been particularly disappointing. Dog walking is a good example. Not surprisingly, studies have reported that people who walk their dogs get more exercise than people who do not own pets, and that dog walkers tend to make friends with people in their neighborhoods. However, these findings do not seem to translate into better health. For example, researchers at Purdue University were able to increase dog walking in an experimental group using email reminders compared to a control group. At the end of the study, however, they found none of the expected differences in blood pressure, blood lipids, or weight loss. British researchers found that dog owners got more exercise than people who did not own pets. Unfortunately, they found no differences in any of the other health or illness measures in the groups. And despite pet industry claims, there is little evidence that walking their dogs causes their owners to shed any pounds. Indeed, a systematic review of twenty-one studies involving over 20,000 participants found no association between dog ownership and obesity.

Will getting a pet, especially a dog, make you live longer? Most studies say no. However, in 2019, a widely publicized review of ten studies was published in the journal *Circulation*. The authors concluded that getting a dog was associated with a 24% reduction in the mortality of pet owners. The problem is that we cannot take these results seriously because of the causal arrow issue—none of these studies took into account differences between dog owners in health-related factors like sex, age, and socioeconomic status. But the same issue of the journal included a high-quality study by a team of epidemiologists at the University of Uppsala that did control for these and other possible confounding variables. Among 300,000 Swedes, dog ownership was associated with an impressive 21% lower risk of death in people who had suffered a heart attack and an 18% reduction among stroke victims. But, unfortunately, these benefits were only accrued by owners of purebred dogs and people who did not have small dogs.

PETS AS HEALTH HAZARDS

On a sunny afternoon in 2016, Diana Rayment of Melbourne, Australia, was taking Bolt, her Rottweiler, for a walk. Suddenly, Bolt spotted a couple of kangaroos and immediately took off after them. Unfortunately, his leash was tangled around Diana's legs. As she fell, Diana's body twisted because of the leash. As she wrote to me in an email:

> This effectively separated the bones of my foot from the bones of my leg.

Luckily, she was soon discovered lying on the ground in pain by a couple of neighborhood kids and rushed to a hospital. The X-rays revealed she had suffered "fully displaced comminuted spiral fractures of both the tibia and fibula"—her lower leg bones. Diana underwent three surgeries over the next year. She now has a permanent limp.

Injuries like Diana's are surprisingly common. In the United States, nearly 90,000 people a year are taken to emergency rooms because of falls caused by their pets. Given the frequency of these types of injuries, I was surprised to find that little research had been done on the topic. So I sent out a message to several human-animal interaction Facebook groups. It said simply, "I'm looking for stories of people injured by tripping over pets." Over the next three days, nearly fifty people sent me tales of woe. Most of these accidents were caused by dogs and a few by cats. Twenty-one involved broken bones—mostly leg, ankle, and foot fractures. But there were also a few shattered pelvises, hips, and wrists. Nearly half of the injuries had long recovery times or resulted in permanent injury. Francis Dauster, a professional dog trainer, tripped while walking her own dog. In addition to breaking her leg, she also wound up with a dental infection that required three root canals. She told me, "Now I go up and down stairs like a ninety-year-old waiting to break a hip. Even if you have walked thousands of dogs, it only takes ONE to knock you off your pins and change your life for a long time."

Pets can be health hazards in other ways as well. A survey by the Opinion Research Corporation found that nearly one in six dog owners reported having an automobile accident or a close call caused by their pet jumping around in their car. And 60% of the pathogens that humans get can be contracted from animals. Among these are roundworms, skin mites, Lyme disease, brucellosis, ringworm, giardia, leptospirosis, E. coli, hookworm, and the aptly named cat scratch fever. When I was twelve, I had a pet baby turtle; everyone did. Who knew that 85% of them carried Salmonella? The FDA banned baby turtle sales in 1975, but now snakes, lizards, and other reptiles are popular pets. Predictably, pet-related Salmonella is on the rise, and 75,000 cases of Salmonella infections each year are contracted from pet reptiles and amphibians living in American homes. Even therapy animals are a potential health hazard. Therapy dogs can acquire and spread MRSA, a serious antibiotic-resistant *Staphylococcus* infection, from patient to patient in hospitals and nursing homes.

WHY ARE HUMANS THE ONLY ANIMAL TO KEEP PETS?

Clearly, the health benefits of pet ownership have been overstated by the media, but there are lots of other reasons for having a companion animal in your life. For instance, pets are fun and our bonds with them can be extraordinarily deep. A survey by the American Animal Hospital Association reported that almost half of the married female pet owners said they got more emotional support from their dogs and cats than from their husbands or kids. In another survey, 82% of dog owners said they worry about their pets when they are away from home, and 47% admitted that they would find it harder to leave their dog for a week than their human partner.

The fact that we can love our pets so deeply begs the evolutionary question of why humans become so attached to members of other species. Darwinism implies that, directly or indirectly, organisms should act in ways that increase their reproductive fitness—that is, their success in passing down their genes to the next generation. But if this is true, why should Joe and his wife, Mary, pay a thousand dollars a month for chemotherapy to keep their aging golden retriever alive for another year or two? Wouldn't their genes be better off if they used the money to pay their children's (or grandchildren's) college tuition bills?

Dan Gilbert of Harvard University claims that every psychologist who puts pen to paper takes a vow to someday write a sentence that begins "The human being is the only animal that . . ."

Here is mine: "The human being is the only animal that keeps and takes care of members of other species for extended periods purely for enjoyment." The question of why we keep pets is an evolutionary mystery, right up there with why humans are the only mammal with complex symbolic language, moral codes, religious beliefs, and the ability to learn to enjoy the burn of red-hot chili peppers.

Oh, I can hear the howls of objections. What about Koko's kitten, you ask, referring to the well-known case of the American Sign Language–trained gorilla who fell in love with a cat? What about Owen, the 600-pound baby hippo who became fast friends with Mzee, a 160-year-old giant tortoise in a Kenyan game preserve? How about Tarra, the Asian elephant, at the Elephant Sanctuary in the hills of Tennessee, whose best friend was a dog named Bella?

You are right. There are scads of examples of long-term attachments between animals of different species. The problem is that nearly all these cases have occurred among captive or semi-captive animals in zoos, wildlife sanctuaries, or research labs. In the wild, our closest relatives on the evolutionary tree, chimpanzees, will occasionally capture a pangolin, duiker, or a squirrel and play with it for a little while. But these interactions don't last long and almost inevitably end in the death of the "playmate." There are, however, three exceptions that I think prove my contention that humans are the only animal to keep pets.

The first case was discovered by Jeanne Shirley, an avid amateur naturalist whose hobby was traveling to the tropics to photograph the wild things. In 2004, while visiting a biological preserve in Brazil, she stumbled across a group of ten bearded capuchin monkeys. She was astounded to see them carrying around a baby marmoset, a much smaller type of monkey. Over the next two hours, she watched and photographed the interactions between the capuchins and the little marmoset. She even saw one of them crack open a pine nut with stone and gently place pieces into the baby's mouth.

She showed the photographs to several primatologists, and they began to systematically study this unique inter-specific friendship. They named the marmoset Fortunata, and they discovered the capuchins treated her just like an infant capuchin. They fed the little monkey and talked to her in capuchinese. They cradled Fortunata, carried her around, and let her ride around on their back like a child rides on a pony. And when they played with Fortunata, they carefully adjusted the force of their movements so they wouldn't injure her. Most importantly, it was not just a transient hookup. The monkeys raised the marmoset from infancy to the age she would have been an adult. One day, about a year after Jeanne first encountered the troop, Fortunata disappeared. No one knows if she went off looking for others of her own kind or was killed by a predator. When the researchers published their observations, they called this a case of "cross-fostering." I call it "pet-keeping."

More recently two other examples of long-term, pet-like adoptions between animals of different species have been reported. In 2018, researchers discovered an Asiatic lioness in the Gir National Park in India breastfeeding a baby leopard along with her own two lion cubs. "The lioness took care of him like one of her own," said Dr. Stotra Chakrabarti, one of the researchers. Unfortunately, the leopard cub died several months after it was adopted. The remarkable case of a female bottle-nose dolphin who raised a baby melon-headed whale lasted much longer. The pair was discovered in 2014 by researchers who were diving in French Polynesia. Like the lion, the dolphin was a lactating mom who was also feeding her own dolphin calf. Unlike the lion, this animal odd couple lasted three years.

Over the last fifty years, biologists and animal behaviorists have spent hundreds of thousands, if not millions, of hours peering through binoculars and crouching in camouflaged blinds observing the behavior of animals in their natural habitat. At this point, they have reported only these three instances of what appears to be pet-keeping in the wild. To me, they are the exceptions that prove my rule that humans are the only animals to keep pets. As shown by the capuchins who raised Fortunata, some nonhuman species could keep pets if they wanted to.

But, they don't.

THE WHEN'S AND WHY'S OF PET-KEEPING

When did this uniquely human phenomenon come to be? We don't have a clue. Among the earliest evidence of pet-keeping was the discovery of a 12,000-year-old gravesite in Israel where a puppy was buried in the arms of its presumed owner, an elderly woman. But it is certainly possible, indeed probable, that our Paleolithic ancestors, like members of some existing tribal societies, captured the occasional parrot or bush pig and brought it home as a pet. Further, it is likely that some lactating Paleolithic women, like the Gir Forest lioness, breastfed the animals they adopted. Indeed, breastfeeding of young animals by human females has been reported nearly worldwide.

The problem is that these very early attachments between humans and animals would not leave any traces in the archeological record. If we were to discover, say, the 40,000-year-old fossil remains of a man cradling a baby monkey, we would not be able to tell if the animal was the dead guy's pet or if the monkey was placed in his grave as a snack in the afterlife.

Lacking solid evidence of when bonds between humans and animals first formed, the best we can do is guess. Individuals that looked exactly like us were living in Africa 100,000 years ago. A real sea change in human thinking occurred perhaps 60,000 or 50,000 years ago as evidenced by an explosion of culture: art, music, weaponry, and tools that were exquisite in form and function. The psychologist Mike Tomasello believes that this revolution in human creativity was fueled by the appearance of a new and sophisticated mental skill—the ability to infer the mental states of other people. Images inscribed on cave walls depicting creatures that were half animal/half human suggest that our ancestors began thinking of animals anthropomorphically 35,000 or 40,000 years ago. James Serpell says the ability to think of animals as you would a person opened the door for taming wild creatures and becoming attached to them. Serpell makes a good case, but without a time machine, we may never know when the first person decided that a ball of fur could be a friend rather than a meal.

IS PET-KEEPING AN EVOLVED ADAPTATION?

As creationists gleefully point out, we Darwinists argue a lot. We don't fight over whether humans evolved from apes or whether the earth is billions of years old. These are facts. Rather, we squabble over the details. The question of why people love pets is emblematic of one of the most contentious debates in evolutionary theory—the issue of adaptation. It is an article of faith among some evolutionary psychologists that if a trait is found in all human cultures, it must have given our ancestors a leg up in the Darwinian gotcha game of Who Can Pass Down the Most Genes. Advocates of this adaptationist school of thought are convinced that the human mind is composed of mental modules that evolved to help our forebears cope with the problems of surviving and reproducing in their Stone Age environments. They believe that the process of natural selection equipped the human brain with specialized modules for skills like learning language, avoiding sex with close relatives, detecting snakes and cheats, and impressing potential mates.

Critics of the adaptationist paradigm argue that aspects of human nature could have evolved even if they have no effect on a person's reproductive success. They think some traits are simply the unintended side effects of characteristics that were adaptations. For example, our bones are white because they are made of calcium, not because women are attracted to men with pallid skeletons. Harvard paleontologist Steven Jay Gould likened nonfunctional biological traits to spandrels, the leftover spaces in a building that architects integrate into the design for aesthetic rather than functional reasons.

As an example of this debate, consider how adaptationists and non-adaptationists explain the evolution of orgasm in human females. Adaptationists (and I was once among their number) have dreamed up a couple of dozen theories to explain why orgasms occur often in human females but rarely if at all in other species. Among their more innovative explanations are that orgasmic contractions suck sperm into the uterus, that orgasms evolved as a signal that helps women tell which men have good genes and which ones are evolutionary losers, and—my favorite—that by making them woozy, the throes of orgasm keep women lying down after sex so sperm don't have to swim uphill to reach the egg. Skeptics of adaptationist thinking scoff at these ideas. They explain orgasm in women as the side effect of the undeniable fact that orgasm has reproductive benefits for men—just like the presence of nipples on men is the nonfunctional by-product of the fact that nipples evolved so female mammals can feed their infants.

The argument over adaptationism also applies to explanations of the human-animal bond. I suspect that most people—including many of my fellow anthrozoologists—want to believe that love of pets is a fundamental attribute of human nature that evolved because it helped our ancestors survive and reproduce. Evolutionary psychologists argue that if a trait is an evolved adaptation, it should be common, widespread, and perhaps, like language, be universal. This is certainly not the case with pets. The anthropologist Donald Brown of the University of California at Santa Barbara compiled a list of nearly 400 human universals. It ranged from thumb-sucking to beliefs about death. "Interest in bioforms" makes the list, but pet-keeping is conspicuously absent.

In many parts of the world, people do not form close bonds with animals. This is particularly true in Africa. My anthropologist friend Nyaga Mwaniki was born in the shadow of Mount Kenya. In his village people never become attached to individual animals. Indeed, there is no word for "pet" in Kiembu, his native language. The villagers use dogs to guard against intruders and chase elephants away from their gardens. But they never allow dogs in the house, they do not think of them as companions, and they would be horrified at the idea of letting one sleep in their bed.

Finally, if pet-keeping is an evolved adaptation, at some point in human history, people bonded to individual animals must have been better at passing on their genes than their less pet-o-philic peers. Your final score in the cosmic game of Darwinism is measured in terms of reproductive success. The fact that your dog or cat makes you happier, healthier, or even helps you live longer is irrelevant. Could living with a pet increase your reproductive fitness? Perhaps girls who grow up with pets are more successful at raising offspring because they learn parenting skills by taking care of the family dog. Or maybe early humans who kept pets were more apt to survive the hard times because they could eat their companion animal. I suppose it is remotely possible that some women are turned on by macho guys with big dogs, and others by kinder and gentler men who demonstrate their sincerity by cuddling puppies. (Recall that Antoine, the handsome Frenchman, got more dates when he had a dog with him.) But I am skeptical that the possible reproductive advantages our ancestors might have accrued by falling in love with, say, a baby monkey outweighed the costs in terms of time and resources.

ARE PETS PARASITES?

But if pet-keeping does not have an evolutionary function, why do we form such close bonds with animals and invest so much money and emotional energy in them? One possibility is that our love of pets is, like, the color of our bones, a nonfunctional evolutionary side effect. Consider Harvard evolutionary psychologist Steven Pinker's theory of music. Pinker is an adaptationist when it comes to language and fear of snakes, but he believes that our love of music is a biologically useless consequence of the way that our brains are wired. Could Pinker's view that music has no adaptive value to humans also apply to our love of pets?

Now consider one of the most common claims Americans make about their pets: "They are my children." Humans are instinctively drawn to animals that remind them of infants, creatures with big eyes, large heads, and soft features. These traits hook our parental instincts and help us pass down our genes by eliciting care for creatures with whom we do share genes: our offspring. But because instincts operate automatically, they can be hijacked. Take, for example, nest parasitism, a reproductive strategy used by dozens of species of birds. A brown-headed cowbird will lay an egg in the nest of an Eastern phoebe. The hapless phoebe will blithely brood the cowbird's egg and then feed the parasitic nestling until it fledges and flies off. Ironically, the phoebe probably gets great emotional satisfaction by raising her faux offspring, not realizing that she has been the victim of a Darwinian sting operation.

Mary Jean and I were victims of this scam. The parasite was none other than our cat, Tilly, and the perp was the savvy mother cat who deposited a baby at our door, never to be seen again. Our yellow lab Tsali had died the year before, but we were not looking for a new pet. When I came home from work one afternoon, Mary Jean greeted me with a big smile; that's when I heard a plaintive *meow* coming from the living room. She had found a kitten under our deck. The little sad sack had the full complement of baby releasers. With those big eyes and soft fur, she was irresistible. It was a done deal.

The idea that the human-animal bond is an evolutionary side effect caused by a misfiring of our parental instincts appeals to me. The problem is that this theory does not explain the large cultural differences that exist in the frequency and styles of pet-keeping. Perhaps a different type of evolution—social evolution—provides a better perspective on why we love our pets than Darwin's evolutionary theory.

IS PET-KEEPING A MENTAL VIRUS?

The high point of my intellectual career may have been in 1979, when I found myself sitting next to the famed evolutionary biologist Richard Dawkins. We were on a bus filled with ethologists headed to the Vancouver Aquarium during a conference. I had just finished his book *The Selfish Gene*, and I was starstruck. The book was important for a lot of reasons, but it was chapter 11 that spun my head around. In it, Dawkins argued that evolutionary change does *not* require genes or even organisms. All you need are replicators—gizmos that can make copies of themselves and have the attributes of longevity, fecundity, fidelity, and the ability to copy accurately. In biological evolution, the gizmos are the molecular spiral staircases that we call genes and that use our bodies to reproduce themselves. Dawkins' insight was that cultural evolution works in the same way. But with cultural change, the gizmos are bits of information that are transmitted by imitation and that use our minds to reproduce themselves. The term *gizmo* does not have much scientific gravitas, so he called his hypothetical units of cultural transmission *memes*, a term he coined because it rhymed with genes and also hearkened up the Greek word for memory.

Memes are everywhere. Some are trivial (nose rings), some are tragic (a short-lived fad in Japan for strangers to commit suicide together by igniting a charcoal brazier in a sealed minivan), some are transcendent (forms of art). Snatches of songs you can't get out of your head are memes. So are cool sneakers, political ideologies, and fake news that goes viral. Our ancestors spread memes by copying each other's behavior. But memes became a much bigger factor in human evolution with the development of symbolic language. Now memes spread around the globe at warp speed via radio and television, and, of course, social media.

The term *meme* is itself an extraordinarily successful meme. When I Googled "meme" ten minutes ago, I got 3.7 trillion hits. The word made it into the Oxford English Dictionary, and last night, a TV commentator described a nasty political smear campaign as a meme. My daughter tells me that the word regularly shows up in youth culture e-mags. But while nerdish techies, edgy hipsters, and a few philosophers have embraced Dawkins' idea, anthropologists, the real experts in cultural evolution, are lukewarm at best. They claim the definition of memes is too loosey-goosey, that—unlike genes—memes are not discrete units. They argue that human cultures march along just fine without the assistance of imaginary replicators.

I am agnostic as to whether memes exist as actual entities. But there is no doubt that ideas and behaviors are contagious, and I find that memes are a useful metaphor for thinking about the role culture plays in our relationships with other species. From a meme's-eye view, pet-keeping is a mental virus spread by imitation. This idea sounds far-fetched, but the evidence for this perverse hypothesis is surprisingly strong.

First, like genes, memes are transmitted from parent to child, but via learning rather than sexual reproduction. Hence, they tend to run in families. Catholic parents usually have Catholic children and circumcised fathers usually have circumcised sons. Similarly, children raised with pets usually grow up to be pet-owning adults. Further, cat kids tend to become cat adults, and dog kids, dog adults.

Second, consistent with the meme hypothesis of pet-keeping (and contrary to the evolutionary adaptation theory), societies differ widely in their attitudes toward and treatment of companion animals. Peter Gray and Sharon Young combed the anthropological record to examine how pet-keeping played out in sixty societies. They found that, for the most part, animals living with humans were not particularly well-treated. Dogs hung around people in fifty-three of the cultures. But they were only fed in twenty-two of them. More importantly, they were played with in only three societies and allowed indoors in only seven. In thirteen of the cultures, dogs were routinely abused, and in eleven they were intentionally killed. In eight cultures, pet dogs were on the menu. Among the species found as pets in these cultures were ostriches, bears, monkeys, prairie dogs, tortoises, wallabies, and armadillos. The anthropologist Jared Diamond was impressed with the differences in attitude toward pets among the peoples he studied in New Guinea and Amazonia. Tribal peoples in the Amazon often kept pets, usually small birds and monkeys. But pet-keeping was almost non-existent in New Guinea. Indeed, while Amazonians often doted on their pet birds, men in New Guinea delighted in torturing small birds by tying them to a string and repeatedly lowering them into a campfire.

Third, pet memes can spread rapidly. For centuries, the pets of choice in Japanese homes were goldfish and caged birds. As the Japanese began to emulate aspects of American culture after World War II, dogs became increasingly popular, and a quarter of Japanese homes now include a dog.

In China, Chairman Mao felt that pet-keeping was a bourgeois affectation, and pets were banned during the Cultural Revolution. The prohibition on pets was lifted in the 1990s, and the number of companion animals rose nearly as rapidly as the number of Kentucky Fried Chicken franchises in Beijing. Presently, 22% of Chinese homes include a pet, and the amount of money the Chinese people spend each year on pets jumped tenfold in the past decade.

Rapidly spreading pet-related memes, however, can be disastrous for animals. On the afternoon of September 3, 1939, thousands of Londoners began killing their dogs and cats. Over the next four days, 400,000 companion animals were euthanized, most of them in veterinary clinics. The pet massacre was triggered by the announcement that Britain had declared war on Germany. That week, approximately 25% of the companion animals in London were killed, a death toll that vastly outnumbered the number of Londoners who died in the Blitz. Having your dog or cat killed suddenly became "a thing." The practice was not mandated by the government. Nor was it a reaction to enemy bombing, which began a year later. Predictably, the slaughter created by the euthanize-your-pet meme was quickly followed by a backlash. By the next spring, many owners were, in the words of historian Philip Ziegler, "regretting the holocaust of pets that occurred at the outbreak of the war."

TILLY-LOVE AND THE MYTH OF SINGLE CAUSATION

Many scientific disputes result from the erroneous belief that if one explanation of a phenomenon is correct, all the others must be wrong. I would like to think there is one correct answer to the question, "Why do humans keep pets?" But, alas, this is not the case.

In explaining animal behavior, ethologists talk about two different types of questions: proximate and ultimate. This distinction also applies to understanding human-animal relationships. Proximate questions focus on the *hows* of behavior—how they work and develop, their underlying neurological and psychological mechanisms. The idea that attachment to pets might be affected by levels of hormones in your bloodstream is a proximate-level explanation. So is the theory that people love pets because they make us feel needed. Ultimate explanations, in contrast, focus on the *whys* of behavior—what their function is, how they evolved, and whether and how they helped our ancestors survive and pass on their genes. The idea that pet-keeping is the result of the misfiring of human maternal instincts is an ultimate-level explanation.

It is important to keep in mind that both proximate and ultimate explanations of pet-keeping can be correct. I like having Tilly around for lots of reasons. I initially found Tilly adorable because her giant eyes and infantile features triggered the parental instincts that helped my forebears pass their genes on to me. This is an ultimate explanation of pet-keeping. But she is fun to play chase-the-laser-pointer-light with, she makes the house feel less empty when Mary Jean is away, and her athleticism sometimes takes my breath away (she can shimmy up a dogwood tree in five seconds flat). She has just plain weaseled her way into my heart. These are proximate explanations of Tilly-love.

Different perspectives on why humans keep pets also reflect the biases of different academic disciplines. Clinical psychologists believe we live with pets because they make us feel loved. Some biologists say pet-keeping is a form of nest parasitism. And some sociologists claim that pets are purely a human social construction; that's why a puppy can be a family member in Kansas, a pariah in Kenya, and lunch in Korea. The bottom line is that our love for pets, the closest of human-animal relationships, is complex and multilayered. Our pets do make us feel needed and they can provide psychological support when times get tough. But they can also be social constructions and parasites.

Don't feel bad, Tilly. I love you even if you are a socially constructed parasite. I am, after all, infected with a mental virus that urges people to bring cats and dogs into their homes and think of them as our children.

Want another crunchy tuna treat, sweetie?

Friends, Foes, and Fashion Statements

THE HUMAN-DOG RELATIONSHIP

> I love playing with dogs. As they age, so many people lose the capacity to play, to have fun and enjoy the moment. I am seventy-eight and dogs keep reminding me how to stay in the moment and to enjoy it. Their smiles, wagging tails, and kisses say it all.
>
> —DR. RUBY R. BENJAMIN, PSYCHOTHERAPIST

> If dogs could talk, it would take all the fun out of owning one.
> —BOB DYLAN, SONGWRITER

"Don't get so close to the fence. He will bite you in the ass."

I moved away from the fence.

"He" was Maverick, an animal whose genetic heritage runs 98% wolf and 2% dog. The speaker was Nancy Brown, owner of Full Moon Farm, a sixteen-acre sanctuary for wolf-dogs (she never calls them "wolves") near Black Mountain, North Carolina. Some were rescued from abusive homes, others were brought to Nancy by wildlife officials or animal control officers. One was given to her by a Maryland couple after their bottle-fed wolf-dog puppy grew up and, in one afternoon, inflicted $10,000 worth of damage to their condo.

Full Moon is ten miles down Highway 9, a two-lane blacktop running through a valley that reminds me of what the Blue Ridge Mountains

looked like thirty years ago, before the gated communities started cropping up. Drive by the Clear Branch Baptist Church, then follow Rock
Creek past a falling-down barn and the volunteer fire department. Take
the left fork, ignore the dead end sign, go a couple more miles, and turn
right on a rough dirt drive you can barely see from the road. You know you
are at the right place when you see the sign that says, THIS PRIVATE PROP-
ERTY IS MAINTAINED FOR THE COMFORT AND SECURITY OF OUR ANIMALS. IF YOU
DON'T LIKE THAT THEN PLEASE GO AWAY.

I am a quarter of a mile from the sanctuary when the animals hear
my car and start to howl—eerie, like an old Western movie. But howls are
intermixed with *arf, arf*—a sound you would expect to hear from golden
retrievers but not from free-range wolves in Montana or northern Italy. By
the time I switch off the engine, I am assaulted by a discordant choir–
seventy animals, skittish, each one shouting "Stranger!" in a mélange of
dialects that reflect their various stations on the twisted evolutionary path
that led canids from the wild to the tame.

Nancy comes out, a cup of coffee in hand, and introduces herself. The
animals are still howling full tilt. She can tell them apart by their voices.
Our conversation is peppered with interruptions: *"Hear that? That's an argument." "That's Aries." "Hi, Guinevere." "Shut UP, Autumn!"* Some of the
wolf-dogs sit up and pay attention when Nancy hollers at them. But most
of them keep pacing. They are nervous around a stranger, tight as the fifth
string on a banjo. They aren't aggressive. Just varying degrees of paranoid.

At first blush, they all look like full-blooded wolves to me. Their coats
range from pure white to brown flecked with black and gray, and they
have an intensity that gets your attention. But Nancy shows me some of
the subtle differences between the high- and low-content animals, and
I start to get it. The ones whose genetic heritage leans more toward dog
have wider faces, thicker ears, stockier legs. They bark more. Blue eyes are
a sure sign that dog blood runs in an animal's veins. The "98% pures" have
the surly James Dean look that wolf-dog groupies love.

Nancy says that the high contents rarely make good companions, that
low-content wolf-dogs are easier to live with. If you get a good low-content

one and train it right, you might be able to put it on a leash, take it for a walk, and play with it. Maybe it won't be constantly on the lookout for a gap in the fence or try to kill your neighbor's cat. In other words, your wolf-dog might make a good pet.

But Nancy cautions me against stereotyping her babies. Even a high-content animal can make a good pet if paired with the right person. She makes her point by entering the enclosure that Maverick (98% wolf) and his pal Mikey share. In a flash, the big animals turn puppy, romping around Nancy, playing with her, cuddling. The chemistry between the three of them is magical. But even with these, her favorites, Nancy has rules. She never lets them get in a position above her head. She never plays tug of war with them. I ask Nancy how many of her animals have the potential to be rehabbed for family living. She looks up and thinks for a minute, mentally ticking them off, and says, "Four." Not good odds out of six dozen.

The legal status of wolf-dogs is murky. In some states, they are banned outright, but in North Carolina, anyone can own a wolf-dog. They are classified as wildlife in Pennsylvania, and you need an exotic wild-life possession permit to keep one on your property. Sandra Piovesan of Salem, Pennsylvania, thought of her nine wolf-dogs as pets rather than exotic wild animals, so she registered them as dogs and treated them like her children. A few weeks after she told one of her neighbors that her wolf-dogs gave her "unqualified love," Sandra's body was found in their enclosure, mauled by her pets.

According to Merritt and Beth Clifton, publishers of the website Animals 24–7, between 1982 and 2020, twenty people in North America were killed by wolf hybrids, and another fifty-seven people were maimed by them. They claim a wolf-dog is sixty times more likely than a German shepherd to maim or kill a child. Wolf-dog fans don't buy it. To them, wolf-dogs are misunderstood, their reputation undeserved.

Driving home from Full Moon, I was keyed up. The animals were magnificent, and I admired Nancy's commitment to them. If not for her, these animals would have been euthanized. Instead, their lives are as good as it gets for creatures whose heritage is an uncomfortable mix of the wild and the tame. But their edginess rubbed off on me. I felt like a journalist who had spent the day riding with an outlaw motorcycle gang.

Unsettled, I went for a walk that evening and ran into my friend Jeanette and her dog Bindi, a mellow little mutt she had rescued from an animal shelter. Bindi looked up at me expectantly, and I reached down and scratched her behind the ears. Suddenly I felt relaxed. It was hard to believe the difference between Bindi and the skittish gray ghosts I had spent the day with boiled down to a few base pairs of DNA.

THE WACKY AND WONDERFUL WORLD OF SHOW DOGS

The best way to observe the seemingly infinite variety of forms that humans have sculpted from wolf genes is to spend a couple of days hanging out at an all-breed conformation show sponsored by the American Kennel Club. The Asheville Kennel Club hosts one of these events every summer, and it attracts 1,500 dogs, their owners, and handlers from around the country. Some fly in but even more show up in Winnebagos and motor homes. The Western North Carolina Agricultural Center's parking area quickly fills with lawn chairs, gas grills, and portable dog pens. Vendors hawk dog paraphernalia—the bite-sized pieces of liver handlers use to make dogs perk up in the show ring; mastodon-sized bones for Great Danes and Saint Bernards; assorted grooming gear, lotions, and shampoo; dietary supplements; bows and hair clips for the little dogs; socks embroidered with breed silhouettes for their owners.

In the main arena, judges, stewards, owners, and casual observers mill around, chatting between events, the owners combing, snipping, and flicking a stray hair on a poodle, wiping slobber off Newfie's face. The dog people look perfectly normal—there are roughly equal numbers of men and women—some fat, some tall—retirees to kids ready for the Junior Showmanship competition. The biggest surprise is how quiet the arena is.

Over a thousand dogs are waiting for their turn in the spotlight, but I don't see any poop and, with the exception of a couple of yappy Chihuahuas, there is little barking. These dogs are professionals.

I wander around, photographing, and asking people about their dogs. Like most of the animal people I have met over the years, their eyes light up when they talk about their show dogs. It's like asking parents about their children. The dog people are easy to talk to, and their enthusiasm for their dogs is infectious.

There is, however, the occasional weirdness at these events. I notice a well-dressed woman, a professional handler, sitting next to the tawny Great Dane she will soon be parading around Ring Four. On the table in front of her is a pile of gummy bears—the chewy rainbow candy kids like. The woman puts a fistful of gummy bears in her mouth and starts to chew. After the candy has been transformed into a slimy mess, she reaches in her mouth, extracts the amorphous glob of sugar and spit, and, nonchalantly, shoves it into the big dog's open maw. Huh?

That afternoon, I mention the gummy bear incident to another handler, a tall blond woman from New Orleans who is getting ready to show a lovely white and black Japanese Chin. "Oh sure," she tells me. "Professional handlers do that all the time. It helps the dogs get to know you. I use chicken." She points to a four-inch chunk of boiled chicken meat tucked into her armband. As she leads the dog, Fred, into the ring, she plops the whole thing into her mouth and tucks it in her cheek, like a wad of Red Man. As the judge approaches her dog, she takes the piece of chicken out of her cheek, waves it an inch in front of Fred's nose, then puts it back into her mouth. Fred perks up.

This day, I am in luck. I run into a woman named Barb Beisel who is grooming a little Havanese. A highly regarded professional dog handler and breeder, she decides to take me under her wing for the next two days. Barb introduces me to a couple of top handlers: Jimmy Moses, who has been showing star-quality German shepherds for decades; and Chris Manelopoulos, whose world revolves around a white standard poodle named Remy. (Seven months later, I watched both of them on national television running their dogs around the ring in the finals at Westminster. Remy took the nonsporting group, but in an upset that made national headlines, he lost Best in Show to the goofy and energetic Uno, the first beagle ever to win.)

Even at the relatively small Asheville show, most of the handlers are professionals. It will cost you $80 to $150 each time a handler takes your dog around the ring. Unless you know what you're doing, the money is well spent. Barb tells me even a really good dog does not have much of a chance of winning without a pro parading it before the judge.

Barb worked in insurance until she took to the dog circuit full-time. She has silver hair and twinkling eyes and looks like the fairy godmother in *Cinderella*. But when she leads a dog into the ring, she is all business. A lot rides on each animal's performance. High rollers in the rarefied world of international dogdom hire Barb to "finish off" their prize pooches. When Barb takes on a new client, a dog she thinks has star power, she first teaches it how to show its stuff in the ring—to look alert, to follow her hand as she waves a morsel of chicken in front of its nose, to patiently put up with judges who poke their withers and peer at their gums. Then she carts the pups around the country until they earn the points needed to be awarded the title of Champion. Barb is always on the move, traveling to dog competitions in her motor home along with an entourage that includes a dozen show-quality terriers, her assistant Marie, and an aging ferret.

Barb loves dogs, but not all breeds, equally. She is partial to the toys—cute, energetic little animals with high-pitched yips and bows in their hair. While Barb was showing me the fine points of grooming a six-pound Yorkie, another handler walked by leading a mastiff the size of a pony, its testicles nearly as big as the Yorkie's head. Barb glanced at the slobbering brute and muttered, "I just don't know how anyone can love a creature like that." I suspect the mastiff's owner may have been thinking the same thing about the lap dog whose lustrous coat Barb was brushing for the twentieth time that afternoon.

THE PATH FROM WOLF TO WHIPPET

How did wild creatures that looked exactly like Nancy Brown's gang of lupine bandits become transformed into animals as different as a giant mastiff and an elfin Yorkie? How did the descendants of gray wolves become the most variable mammal on earth? A mastiff can weigh 200 pounds. The smallest dog in the world is a four-inch-high Chihuahua named Milly who weighs almost one pound. This difference in size is proportionally larger than that between a full-grown African elephant and me. Molecular biologists Heidi Parker and Elaine Ostrander of the National Human Genome Research Institute say that the domestication of dogs is the most complex and extensive genetic experiment in human history. Remarkably, this morphological miracle took place in the blink of an evolutionary eye.

Charles Darwin thought the modern dog was a mix of coyote, wolf, and jackal. He was wrong. If we know anything for sure about canine evolution, it is that the ancestors of the dog living in your house—whether Pekingese or Rottweiler—were gray wolves. Dogs and wolves are so genetically similar that the American Society of Mammalogists classifies the domestic dog as a subspecies of the wolf. It makes sense that the wolf was the first domesticated animal. Like us, they are social, active during the day and good at figuring out who is the boss. Hooking up with odd-looking and hairless two-legged creatures was a strategy that worked out well for *Canus lupus familiaris*; about a billion dogs are running around the planet right now, compared to a couple of hundred thousand gray wolves. But when, where, and why did our ancestors first invite a predator into their lives?

First, when. The answer is not all that long ago. Nearly every dog owner I know has a slew of pictures of their pets around the house. If our Stone Age ancestors cared as much for dogs as we do, they probably would have made pictures of them too. But they didn't. Paleolithic cave paintings of animals go back 45,000 years—wild pigs, buffalo, horses, mammoths, ibex, rhinoceroses, bears, lions, deer, and even a few fish and birds. But not a single creature that looks like a dog or wolf appears early in the Stone Age bestiary. To learn how dogs and humans came to throw in their lots together, we must turn to bones and genes.

Members of the genus *Homo* have lived alongside wolves for 400,000 years, but there are no signs that wolves and early humans were on friendly terms. Evidence of relationships between man and wolfish-looking dogs in the fossil records show up as early as 14,000 years ago. Domestication changes a species. First, they get smaller. Juliet Clutton-Brock of the Natural History Museum in London believed this was an adaptation to malnutrition during pregnancy and natural selection for larger litters. The earliest dogs had smaller jaws with more crowded teeth, wider faces, and shorter snouts than wolves. Like most domesticated animals, they also had smaller brains than their progenitors. In short, the original dogs looked a bit like puppified wolves.

Dogs quickly became fixtures of late Stone Age cultures in Europe and Asia. New evidence suggests humans were selectively breeding dogs in the Arctic 9,500 years ago. But most of the three or four hundred current breeds trace back less than 150 years to when the first kennel clubs were organized in England and the United States. By 10,000 years ago, dogs were romping around the New World, having accompanied humans who made their way across the land bridge from Siberia to North America. It took only 4,000 years for dogs to make the long journey from Alaska to Patagonia.

The tools of the archeologist's trade are spades, dental picks, and dusty hiking boots, but molecular biologists trying to ferret out the mysteries of canine evolution wear lab coats and spend their days listening to the hum of DNA sequencers. The raw materials for their research are not fossilized bits of jaw and teeth, but fragile strands of genetic material, mitochondrial DNA that has been passed from mother to child over thousands of generations. Mitochondrial DNA is different from the nuclear DNA that makes up the genetic instructions you bestow on your kids when sperm meets egg. Mitochondrial DNA floats in a cell's cytoplasmic goo, converting organic material into energy. Most importantly for evolutionary biologists, all your mitochondrial DNA comes from your mother. Unlike regular DNA, which is randomly shuffled during sex, ancestral lines of mitochondrial DNA remain unchanged from generation to generation, except for the occasional mutation.

These mutations are important because geneticists can use the rate of mutational change in mitochondrial DNA as a kind of molecular clock that indicates when, where, and from what a species has emerged. In 1997, an article appeared in the journal *Science* that shook up the world of people who care about things like when dogs made the great leap from wolves. A group of researchers led by Robert Wayne of UCLA analyzed the mitochondrial DNA of sixty-seven dog breeds as well as the DNA of wolves, coyotes, and jackals. Based on their interpretation of the molecular clock, they concluded that dogs emerged from gray wolves between 100,000 and 135,000 years ago, five to ten times earlier than the accepted chronology of the dog-wolf split. That is a huge discrepancy. What's going on? The accuracy of a molecular clock, just like an alarm clock, depends on when you set it, and the UCLA researchers may have been overly generous when calibrating theirs. Most molecular biologists now believe that dogs began to diverge from wolves between 15,000 to 50,000 years ago, dates that jive better with the story the bones tell.

WERE THE FIRST DOGS DUMPSTER DIVERS AND DOO-DOO EATERS?

Anthrozoologists disagree about how humans and dogs came to throw in their lots together. The traditional theory goes like this: Fred Flintstone, Stone Age hunter, stumbles upon a wolf den one afternoon while foraging for dinner and brings a baby wolf back to the cave to cook for the evening meal. But just as Fred was about to toss little Lobo on the campfire, his wife, Wilma, looks deeply into those adorable puppy eyes and her maternal instincts kick in. She snatches the puppy from Fred and dotes on it as if it were her child, maybe even suckling the pup. The wolf's fierce nature is soon tamed, in Bill Monroe's words, "by a good woman's love." Voila, Lobo becomes the first domesticated companion animal, and soon produces puppies that, over time, come to look and act like Rin Tin Tin.

The late biologist and dog-sled racer Ray Coppinger dismissed this fantasy as the "Pinocchio Hypothesis." He thought converting full-blown wolves into dogs was about as likely as the fairy godmother in *Pinocchio* waving her magic wand over the wooden puppet and turning it into a flesh-and-blood boy. The trouble with the wolf-into-puppy-dog theory, said Coppinger, was that pure-blooded wolves were essentially untamable. That's why you didn't see them in circus animal acts. Coppinger admitted that wolves can learn a few tricks and that some will become habituated to walking on a leash, but he feels this is a far cry from being truly tame. Other researchers support Coppinger's view. Even hand-reared wolves do not become attached to their caretakers the way dogs do. They are more likely to growl and literally bite the hand that feeds them.

But if humans did not tame wolves, how did they become our pals? They may have tamed themselves. Ray Coppinger traced the emergence of dogs to when our ancestors began to develop agriculture and settle into permanent villages. Settlements generate piles of refuse, potential gold mines for opportunistic scavengers. Wolves who were less nervous around humans would be more efficient garbage-pickers, just like the dogs-gone-wild you can see today hanging around garbage dumps on the outskirts of Lagos or Mexico City. These animals would have been better nourished than their warier competitors, hence bear more offspring—puppies that would share the genes for being less fearful around humans. Over generations, these tamer animals would have access to more food and eventually adapt to living around people. This self-domestication process would set the stage for unconscious selective breeding by humans for functional traits like barking at strangers, cleaning up after-dinner scraps, and, most importantly, hunting.

Not everyone, however, buys the self-domestication/garbage dump hypothesis of canine evolution. Some researchers argue that the path from wolf to dog had to have been a two-way street. And as the anthrozo-ologist John Bradshaw points out, the onset of the separation of dogs be-gan well before the invention of agriculture which only dates back about 12,000 years. While it is decidedly distasteful, he suggests a different food source that could have motivated early proto-dogs to hang around our ancestor—human feces.

Many modern dogs are coprophagous. This means they like to eat poop. Our dog Molly was one of them. She loved the fresh dung depos-ited by the cows that lived in the pasture behind our house. In parts of the world that lack flush toilets, dogs readily consume human feces. Re-searchers studying wolves and feral dogs in the Bale Mountains of Ethio-pia found that 20% of the dogs' meals consisted of human poop, making it the second most popular item on the canine menu. James Butler and his colleagues in Zimbabwe found that more than half of the meals eaten by free-range dogs included feces. Further, as dog food goes, it is surprisingly nutritious. Their analysis revealed that human poop was "comparable to the upper range of energy content for mammal tissue, vegetables, and fruit."

HUNTING LIZARDS WITH DOGS

Hunting played a critical role in the evolution of the human-dog relation-ship early on and, in some places, continues today. The anthropologist Jeremy Koster has studied how indigenous subsistence hunters use dogs to locate and chase prey in the jungles of Nicaragua. Depending on the type of prey, taking a dog could nearly double the productivity of hunt-ing trips. He also noted that, except for puppies, these hunters were not particularly fond of their canine hunting partners, though they did value them as watchdogs.

I once got a glimpse of how early humans may have used the first dogs to chase down small game. My graduate school friend Bev Dugan spent several years in the jungles of Panama figuring out the sexual strategies of green iguanas. She had a problem: All iguanas look more or less like reptilian versions of Keith Richards. If her project was going to succeed, she needed to mark the animals to tell them apart. But how do you catch creatures that live high in the canopy of a tropical forest?

That's where dogs come in.

Another researcher introduced Bev to Pifo and Caesar, local men who had perfected the art of hunting big iguanas. When I visited her research site, they invited me to tag along on one of their forays. Pifo's dog was a brownish mongrel, medium-sized, with short hair, a curly tail, and perky ears—the classic profile of village dogs who are allowed to breed freely. Ray Coppinger called these "natural breeds." By 5,000 years ago, village dogs were, aside from humans, the most widely distributed mammal on earth. Even today, 85% of the dogs, to some extent, run free. Village dogs often resemble an Australian dingo, and you find them from Tanzania and Israel to the coast of South Carolina. The good news is that village dogs get to mostly call their own shots—how they spend their day and who they mate with. The bad news is that 80% or 90% of puppies will not make it to their first birthday.

An iguana hunt begins when you spot a lizard on a limb basking in the sun. The acrobatic Caesar shimmies up the tree trunk and gingerly makes his way out on the branch. The dog, tense, stares up toward the lizard. Caesar violently shakes the branch until the iguana bails and plunges straight down sixty or seventy feet into the underbrush. In a flash, the dog is after it. There is no way a human could chase the lizard down, but thanks to Pifo's dog, *no problema*.

Pifo and Bev take off after the dog. Clumsily, I do my best to keep up. The dog quickly corners the lizard. The trick for Bev is to get to the iguana before the dog does any damage to it. Male iguanas can be six feet long. They will whip you with their tails and bloody your arms with their claws. You do not want to get bitten by one. Bev usually wins these woman-versus-lizard wrestling matches. Once she has the iguana under control, she weighs it, checks its reproductive status, marks it, and lets it go.

It is an efficient way to hunt. With the aid of the dog, they can capture ten or twenty lizards a day. The day I spent chasing lizards, Pifo and Caesar were catching iguanas for Bev's research project, but most of the time they hunt for meat. Once they catch an iguana, Pifo and Caesar immobilize it by extracting a tendon from the iguana's forefeet and they use it to tie its legs together behind its back. Later, they kill the lizard by inserting a long spiky thorn straight through the animal's brain and into the spinal cord.

Coppinger argued that pure-blooded wolves are too wild for this sort of cooperation between man and beast. After spending time with Nancy's wolf-dogs, I am inclined to agree. But the most convincing support for Coppinger's theory of dog domestication comes from an extraordinary genetic experiment with foxes.

FROM WILD TO TAME: HOW TO MAKE FOXES FRIENDLY

In the mid-1950s, a geneticist named Dmitry Belyaev found himself on the losing end of a scientific dispute and was fired from his position at Moscow's Central Research Laboratory of Fur Breeding. Belyaev wound up as head of the Institute for Cytology and Genetics in Novosibirsk, Siberia, where, in 1959, he began a remarkable experiment on Arctic foxes that has been going on for sixty years and has involved over 50,000 animals. This line of research changed the way scientists think about the transformation of wolves into dogs.

By selectively breeding foxes for tameness around humans, Belyaev wanted to create a strain of foxes that would be easier for fur farmers to work with. He tested fox kits for their reactions to people and then crossbred the tamest 10% of males and females. By the tenth generation, 20% of foxes were tame, and after forty generations, 80% of fox kits were tame. The selectively bred tame foxes would lick your face; the foxes in the control line would rip it off.

The behavioral ecologist Lee Dugatkin has been involved with the project for years. In his book with Lyudmila Trut, *How to Tame a Fox (and Build a Dog): Visionary Scientists and a Siberian Tale of Jump-Started Evolution*, they describe the changes that came with tameness. In just fifteen generations, stress hormone levels in the tame strain dropped by half. The size of their adrenal glands decreased, and their levels of serotonin (a "happiness" hormone) went up. Compared to normal foxes, their neurons produced more of the brain's natural antidepressants. Recent studies have shown that selection for tameness changed the expression of more than 100 genes in the prefrontal cortex of tame animals. No wonder they were mellow.

So, foxes selected for tameness became tame. Big deal. Here is the important part: an unintended by-product of selection for tameness was that the foxes began to look and act like dogs. Over the generations, their ears became floppy and their tails curly. Their coats began to show traces of brown (the probable color of early dogs), and some of them developed the white patches you sometimes see in the faces of dogs. The foxes' faces became shorter, wider, cuter. The animals showed decidedly un-foxlike behaviors. They wagged their tails when people greeted them, and the foxes would whine for human attention—just like our dog Molly.

CAN DOGS READ OUR MINDS?

When I would hear Doc Watson, the great flat picker from Deep Gap, North Carolina, lay out just the right notes in a fiddle tune guitar break, a tingle would run from my left shoulder and up the side of my neck and I would get goosebumps. The Hawaiians call this "chicken skin music." As a scientist, I read a lot of articles in academic journals. Most of them put me to sleep, but now and then, one of them gives me the chicken skin. It happened in 2002 when one of my colleagues handed me an unpublished manuscript that his old college roommate, Mike Tomasello, had sent him. The paper, which soon appeared in the journal *Science*, compared the ability of wolves and dogs to read human gestures. The lead researcher was a twenty-six-year-old graduate student named Brian Hare, and his study jump-started the now-thriving field called canine cognition.

To understand the importance of Hare's research, you have to know something about chimpanzees. Along with bonobos, they are our closest, smartest, and most socially savvy relatives. Like other nonhuman primates, however, chimps are hopeless when it comes to following human signals to find food. Even chimpanzees that are hand-raised by people are not able to use gazing and finger-pointing to locate food. Given that big-brained chimps are so lousy at this task, you would predict that wolves, which have much smaller brains than apes, would also fail the point-gaze test. You would be right. You would also, of course, think that domestic dogs would be dumber than wolves as their brains are roughly 25% smaller. Wrong.

Hare, now the director of the Duke Canine Cognition Center, developed a simple choice experiment to show that dogs are naturals at understanding human signals. He hid food under one of two bowls. A research assistant would tap, point, and gaze at the bowl with the hidden food. As expected, the wolves consistently failed at the task. They randomly guessed the left or right bowl and were right 50% of the time. The dogs did much better. When given all three cues, the dogs made the right choice 85% of the time.

Is the superior performance of dogs the result of nature or nurture? Hare argued nature. He proposed that evolution endowed dogs with a special ability to read humans. The origins of the differences between dogs and wolves have become a matter of hot debate. Canine researcher Clive Wynne argued that Hare was wrong. He attributed the differences to socialization. Subsequent research has shown that, as is often the case, the ability of canids to understand human gestures is a matter of both nature and nurture. About the same time Hare was conducting his studies, Ádám Miklósi's Family Dog Project in Budapest was also comparing the behavior of dogs and wolves. They have found wolf pups can be trained to attend to human signals, but it takes a wolf ten times longer to achieve a similar level of success as a dog and requires vastly more socialization. Wynne and his colleagues subsequently reported that dogs raised in animal shelters with almost no positive contact with people are initially hopeless at understanding human gestures. They can, however, quickly learn to use human points to find food. Researchers recently traced differences in the friendliness of dogs and wolves to changes in just a handful of genes with names like GTF2I, GTF2IRDI, and, my favorite, BICF2G630798942.

THE GOLDEN AGE OF CANINE RESEARCH

Brian Hare's and Ádám Miklósi's studies ushered in a golden age of research on canine behavior. Dog research centers soon began popping up at major universities. As Yale University's Paul Bloom put it, in the study of animal cognition, dogs have become the new chimpanzee. He is right. In 2002, the year Brian Hare's paper appeared, eighteen research articles were published in scientific journals on dog cognition. In 2018, 180 papers were published on the topic. Unexpected examples of canine brainpower appear every month in journals like *Science, Animal Behavior, Journal of Comparative Psychology,* and *Animal Cognition.* Examples abound. The late John Pilley, a retired psychology professor from Wofford College, taught his border collie Chaser to understand over 1,000 English words for objects. Hungarian researchers have found dogs (and even a cat) could learn to imitate novel human behaviors in a canine version of Simon Says. Pioneered by neuroscientist Gregory Berns at Emory University, researchers are using functional MRI, a brain imaging technique, to investigate what goes on inside of dogs' heads. They have found, for example, that whether a dog will make a good assistance animal can be predicted by individual differences in neural activity in two areas of the canine brain—the *caudate nucleus* and the *amygdala.*

As is often the case in science, disputes exist. All the dog behavior researchers I know concur that dogs are special. They do not agree on what makes them unique. This is illustrated by the bestsellers written by a couple of top researchers with different perspectives on the evolution of the human-dog relationship. As the title of their book implies, *The Genius of Dogs: How Dogs Are Smarter than You Think*, Brian Hare and Vanessa Woods argue that dogs, in their own way, are particularly smart. They think dogs possess a set of cognitive abilities that enable them to understand the human mind. Clive Wynne, on the other hand, is not impressed with the intelligence of man's best friend. Indeed, he refers to his dog, a mutt named Xephos, as "a loveable idiot." The title of Wynne's book reflects his perspective on dog uniqueness—*Dog Is Love: Why and How Your Dog Loves You*. Based on experiments in his lab, Wynne concluded that the reason dogs have become the most successful nonhuman mammal on earth is an enormous increase in their capacity to rapidly form emotional attachments with other creatures, including humans.

Some of the most innovative research in canine science focuses on a hot topic—breed differences. There are over 400 breeds of dogs, and most of them were intentionally developed by humans within the last 150 years. Behavioral differences between breeds are obvious. Everyone knows Labs are more inclined to retrieve a Frisbee than, say, a French bulldog. But not all Labs are alike. Our black Lab Molly would play fetch all day, but given her druthers, Tsali, our yellow Lab, would much prefer lying in front of the fireplace. Indeed, the mantra among some canine scientists is "differences within breeds are bigger than differences between breeds.

In recent years, dog researchers have turned to "citizen science" to investigate the origins of breed differences. James Serpell and his colleagues at the Center for the Interaction of Animals and Society developed a web-based survey in which owners could log on and rate behaviors in their dogs. The fourteen traits ranged from dog rivalry to chasing to being overly sensitive to touch. Their Canine Behavioral Assessment and Research Questionnaire (C-BARQ for short) has now been completed by more than 60,000 dog owners, and the data used in over 120 published studies. One of Serpell's first studies examined breed differences

sylvania researchers evaluated the propensity of 4,000 dogs from thirty-three breeds to bite strangers, turn on their owners, and pick fights with other dogs. Surprisingly, pit bulls were not the breed most likely to attack people—Chihuahuas and dachshunds were. Indeed, pit bulls ranked in the middle of the pack, about the same as poodles.

Breed differences in rates of serious attacks on humans raise one of the most contentious issues in our relationships with animals: Are breeds like pit bulls so inherently dangerous that it should be illegal to own one? In her book, *Pit Bull: The Battle Over an American Icon*, Bronwen Dickey traces the history of the demonization of the breed and why it has become the focus of one of the most divisive of companion animal issues. A hundred years ago, pit bulls were considered a model of courage and loyalty, an excellent family pet. A pit bull named Petey was a central character in *The Little Rascals* films. During World War I, pit bulls were featured on military recruiting posters. The image of the breed radically changed in the 1970s and 1980s when pit bulls became enmeshed in racial politics associated with dogfighting, drugs, and crime. In 1987, the executive director of the Arizona Humane Society said the pit bull "is aggressive to anything that moves. . . . In short, it is a four-legged fighting machine and killer." And *U.S. News & World Report* called it "the most dangerous dog in America."

In the late 1980s, in response to sensationalized media attention to pit bull attacks, municipalities in the United States began to enact breed specific legislation that bans or severely restricts the possession of breeds tagged as inherently dangerous; Denver, Colorado, was an early adopter. The reason was a tragic incident. On a late spring day in 1989, Wilber Billingsly, a fifty-eight-year-old evangelical minister, was walking to the store to pick up a couple of things for his wife when an escaped pit bull named Tate grabbed him by the leg and began to chew. Billingsly's neighbor, Normal Cable, heard him wailing, ran out of the house, and started beating on the dog with a two-by-four. When Tate kept gnawing on the pastor's leg, Cable ran back into his house, got his shotgun, and killed the dog. Wilber Billingsly was bitten more than seventy times and both his legs were broken. Tate's owner was aghast. "We never had any problems with him before," he told a newspaper reporter.

In the wake of the attack, the city council enacted a ban on any dog that looked like an American pit bull terrier, an American Staffordshire terrier, or a Staffordshire bull terrier. In a contentious city council meeting, the head of animal control reported that in the previous year, eighty-one people in Denver had been attacked by pit bulls. Over the next two decades, hundreds of towns and counties in the United States passed similar ordinances, as did many nations, including Great Britain, Denmark, Ireland, Spain, and New Zealand.

But times change, and, in some circles, the pit bull's image has been rehabilitated. Indeed, by two to one, the citizens of Denver voted in 2020 to repeal the city's blanket breed ban. Pit bulls in Denver now have a three-year probationary period in which they are given a breed-restricted license. After that, they are treated like any other dog.

This change in attitude has been propelled by several factors. The mantra of bull supporters—"Ban the deed, not the breed!"—has gained traction. Nearly all heavy-hitter animal welfare organizations including the American Veterinary Medical Association, the American Society for the Prevention of Cruelty to Animals, the Humane Society of the United States, and American Humane have come out against breed-specific legislation. PETA is the exception. They argue that pit bulls are more likely than other breeds to be mistreated and euthanized in shelters. PETA wants pit bulls gone.

Breed ban opponents also argue there is no evidence that anti–pit bull laws have reduced dog-related deaths and maiming. They correctly claim that the term "pit bull" has become a catch-all term that is so broad it is meaningless. Even experts are terrible at determining whether a dog is a pit bull. A 2015 study examined the ability of animal shelter workers to accurately identify pit bulls by comparing their breed determinations of dogs with the results of DNA profiles. The misidentification rate was stunning. One-third of the dogs that the pros thought were pit bulls did not carry any pit bull DNA. And the shelter staff missed 20% of dogs that did contain pit bull genes.

The New Yorker writer Malcolm Gladwell argues that pit bull bans are the canine equivalents of racial profiling. The vast majority of pit bulls, he points out, will never bite or attack a person. He writes:

> The kinds of dogs that kill people change over time, because the popularity of certain breeds changes over time. . . . When we have more problems with pit bulls, it's not necessarily a sign that pit bulls are more dangerous than other dogs. It could just be a sign that pit bulls have become more numerous.

The rise and fall of Rottweilers certainly supports Gladwell's hypothesis. In 1993, Rottweilers temporarily overtook pit bulls as America's most dangerous dog. In the 1970s, most people had never heard of a Rottweiler. The breed ranked fortieth in popularity in the United States, with only 3,000 new puppies registered a year with the American Kennel Club. In contrast, with 60,000 puppy registrations a year, German shepherds were among the most loved dogs in America. But in the mid-1980s, while German shepherd registrations stagnated, Rottweilers suddenly became hot. Between 1980 and 1993, Rottie registrations soared to over 100,000 new puppies a year. They were suddenly the second most popular breed in the United States, and a million registered Rottweilers were living in American homes.

The surge in demand for Rottweilers came at a cost. Between 1979 and 1990, Rottweilers killed six people in the United States compared to thirteen fatalities by German shepherds. Abruptly, however, the gruesome statistics reversed. Over the next eight years, German shepherds killed four people while Rottweilers killed thirty-three. As Gladwell would have predicted, the spike in deaths was attributable to their increased numbers. Even if a breed's temperament is unchanged, more dogs translate into more dog attacks. And the bigger the dog, the more damage they can cause. But the breed soon became the victim of its own popularity. The increased mayhem meant more negative publicity, and insurance companies began to cancel the policies of Rottweiler owners. Over the next decade, Rottweiler registrations tanked, falling from 100,000 puppies a year to fewer than 20,000. As the breed's popularity went down, so did the death toll. Between 2016 and 2019, eight Americans were killed by Rottweilers and eight by German shepherds.

OODLES OF POODLES: WHY DO DOG BREEDS SUDDENLY BECOME POPULAR?

The rapid rise and fall in the popularity of dog breeds raises the more general question of what fuels changes in human cultures. Take Crocs, those ugly squishy plastic shoes that, as *Slate* columnist Meghan O'Rourke noted, spread around the globe as fast as a Paris Hilton sex tape. Did Crocs become popular because they were in some objective sense "superior" to other shoes—because they were cheaper, more comfortable and better for your back? Or were they simply a fad—a culturally transmitted mental virus that swept across the American landscape?

We can ask the same question about the popularity of Rottweilers. After all, they are expensive to feed, have to be delivered by C-section, and are prone to hip dysplasia, diabetes, cataracts, and Addison's Disease. Theirs was not, however, the biggest fad in American canine history. That distinction belongs to poodles. Between 1946 and 2007, five and a half million poodle puppies were registered with the AKC, two million more than the runner-up, Labrador retrievers. Poodle popularity peaked in 1969 when nearly a third of all new purebred registrations were for poodles and the AKC hired a full-time staffer just to handle the glut of poodle paperwork. The ascendency of the breed was fast; yearly registrations increased 12,000% between 1949 and 1969. The poodle craze was not limited to dogs. Poodle skirts, usually white or pink and embellished with the silhouette of a French poodle, became *de rigeur* for the bobby-sox crowd. Today, they are hot items on eBay.

Consider now the English toy spaniel. While poodles were invading postwar American culture like army ants on the march, new toy spaniel registrations went from a measly 123 puppies in 1949 to an even lower forty-five in 1969. These dogs are small and cute. They look like Lady in *Lady and the Tramp*. According to the official AKC breed description, they are "bright, loving, willing to please." What could be better than that? Why then, in 2020, were they the 136th most popular breed? The difference in the popularity of these two breeds raises an issue that transcends our taste in pets. Why do some sneaker styles, foods, songs, colors, books, religions—you name it—catch on while others languish in obscurity?

One possibility is that, like genetic mutations that increase an organism's reproductive fitness, some cultural innovations become spectacularly successful because they are better than the competition at doing something. Screwdrivers, safety razors, and pop-top beer cans come to mind. But others, such as transient enthusiasms for disco music or fondue pots, seem more like nonfunctional variations on a theme. Did poodles come to dominate the purebred dog niche in American homes because they were smarter or more obedient or easier to fall in love with than the loser breeds? Or, like the Kardashians, were they just famous for being famous?

I became interested in this question when I was asked to write an article on the evolutionary psychology of human-animal relationships and I decided to include a section on cultural evolution. As I was trolling the Internet for examples of rapid changes in attitudes toward animals, I stumbled on an American Kennel Club website. It listed the numbers of puppy registrations for each breed for the previous three years. Scanning the columns of figures, I noticed that the number of Dalmatian puppies had dropped precipitously. I contacted a Dalmatian breeder who was on the AKC Board of Directors and asked if she could arrange for the organization to give me the Dalmatian puppy registration statistics over a longer period of time. A few weeks later, a thick package from the AKC's Manhattan headquarters showed up at my office. It was the mother lode—sixty years of registration numbers for every AKC breed. Forty-eight million dogs.

That's the good news. The bad news was that I had to make sense of the largest data set in the history of psychology. Not knowing what else to do, I made graphs charting the growth of each breed. I found that since the end of World War II, only four breeds have been America's most popular dog. They are, in order—cocker spaniels, beagles, poodles, cocker spaniels (yes, again), and Labrador retrievers. These breeds have taken very different routes to capturing the hearts of American pet owners. The poodle's path to popularity was meteoric, but Labs took a slow and steady road to number one, increasing at a rate of about 10% a year over four decades. The graphs for cocker spaniels and beagles were wavy.

The graphs also revealed that most dog breeds never weasel their ways into our homes. How many of your friends own an otterhound, a breed that maxed out in 1993 at seventy puppies? Or a Harrier, whose registrations have hovered between six and forty since 1934? I soon became obsessed with deciphering America's shifting preferences for canine companions. At night, I dreamed about demographic curves. During the day, I spent way too much time flipping through breed growth charts, boring my friends with dog talk, and thinking about obscure breeds with weird names—keeshonden, schipperkes, pulik.

The fact is that I was getting nowhere.

POPULAR DOG BREEDS ARE LIKE POPULAR BABY NAMES

My lucky break came when I ran across an article in the journal *Biology Letters* written by two young investigators. Matt Hahn was a biology graduate student at Duke University and Alex Bentley was a post-doc in anthropology at University College London. Their article was about names that people give their babies.

Matt and Alex were from different disciplines but they both studied how dumb luck affects evolutionary change—in Alex's case, how cultures change, and in Matt's, how genes evolve. They hypothesized that most shifts in human culture—from designs engraved on Neolithic pottery to country music hits—are attributable to the fact that humans are inveterate copycats. Baby names offered an ideal way to test their ideas. That's because the Social Security Administration maintains a website with the frequencies of the thousand most common first names in the United States going back to the 1880s.

Matt and Alex downloaded the names of a gazillion babies and went to work. Using sophisticated computer models developed to study changes in gene frequencies, they discovered that continual shifts in the popularity of first names are explained by a theory developed to explain a mechanism of evolutionary change called random drift. The basic idea is startlingly simple: Styles change because people unconsciously copy each other. Occasionally, someone invents a new baby name or food or dog. Whether it gets copied and becomes popular is largely a matter of random chance. In matters of taste, we tend to follow the herd.

I did not understand the mathematics that Matt and Alex used to prove that baby names become popular by dumb luck, but I got the gist of their argument. It dawned on me that shifts in trendy puppies might offer another way to test their ideas. I emailed them and attached a couple of my dog graphs. For good measure, I added:

By the way, my data set consists of forty-eight million puppies. Interested?

Their response was immediate and enthusiastic:

Yes! Send us the file.

Three days later, their analysis revealed that just like baby names, dog breeds get popular by a throw of the cosmic dice. They found that tastes in dog breeds follow a type of statistical distribution that mathematicians call *power laws*. In human societies, these show up in situations in which large numbers of people influence each other. Power law graphs are elegant in the way that a Brancusi statuette is. The line swoops down from the top left, dwindling gradually to right. Malcolm Gladwell, *The New Yorker* writer who has made a career of pointing out the real-world implications of arcane findings in the behavioral sciences, compared power law graphs to a hockey stick lying on the floor, blade pointing up.

The message of the power law is that, whether we are talking best-selling books, citations of scientific papers, music downloads, webpage hits, baby names, or dog breeds, 20% of the available options attract 80% of the attention, a phenomenon economists call the 80:20 rule. After the first couple of choices, popularity nosedives, dribbling closer and closer to zero. That's why in business circles, power laws have come to be referred to as "the long tail."

Here's how it works with dogs. In 2007, 81% of puppy registrations came from the top thirty-one breeds. This left the other 125 breeds to fight over the popularity crumbs. The bottom fifty breeds together attracted only 1% of all new puppy registrations. Labrador retrievers, the most popular breed, generated 9,000 times more puppy registrations than English fox hounds, the least popular breed. That's the inevitability of the long tail.

According to the random drift hypothesis, fashionable cultural tidbits are constantly changing because new innovations occasionally pop up. A man, for example, invents an ugly plastic slip-on shoe he names after crocodiles, or someone develops a new dog breed by crossing a minia-ture schnauzer with a Yorkie (official name: Snorkie). Most new ideas are flops, but every now and then one will catch on. An implication to this throw-of-the-dice view of cultural change is that, for all practical pur-poses, it is impossible to predict the next big thing—no matter if it is a shoe style, a baby name, a pop song, or a dog breed.

Albert Schweitzer once quipped, "When people are free to do as they please, they usually imitate each other." His observation applies to cycles in the popularity of dog breeds. Fueled by our penchant for unconscious imitation, our tastes in dogs seem to reflect the same mob psychology that leads people to think nose rings are sexy. My colleagues and I have now published nearly a dozen papers on the social dynamics of dog breed popularity. For the most part, they support the idea that our choices in pets reflect the influence of fashion rather than a rational evaluation of a breed's characteristics. Here are a few of our findings.

- Breed popularity is unrelated to canine behavior. Dog breeds that score high on traits like hyperactivity, aggressiveness to people and other dogs, chasing, fearfulness, and low trainability become just as popular as sweet breeds that are easy to live with. (However, the more popular breeds tend to have more genetic disorders.)

- Dog breeds are socially contagious. Take the Great Irish Setter Epidemic. In the 1950s, Irish setter registrations hovered between 2,000 and 3,000 a year. In 1962, they reached the tipping point. Registrations suddenly took off, jumping to over 60,000 in 1974, an increase of 2,300%. Then, as suddenly as it rose, their popularity plummeted. By the end of the burnout stage, Irish setter registrations were only 5% of what they were at the peak of their popularity. The graph of their rise and fall is perfectly symmetrical, and, from start to finish, their fifteen minutes of fame lasted exactly twenty-five years. A dozen other breeds, including Dobermans, Old English sheepdogs, and Saint Bernards, have shown the same pattern.

- The Dog Movie Star Effect is real. For example, the 1959 release of the Disney movie *The Shaggy Dog* was followed by a hundredfold increase in Old English sheepdog registrations. However, dog movies have less influence on a breed's popularity now than they did fifty years ago.

- Contrary to conventional wisdom, winning the Westminster Dog Show does not produce surges in demand for a breed.

THE RISE AND FALL OF THE PEDIGREED DOG

Epidemics in the popularity of these breeds illustrate the huge influence that cultural whims have on our relationships with animals. Presently we seem to be in the burnout phase of another canine fad—the desire of Americans to own purebred dogs.

The transformation of American dogs from pals and working partners to fashion statements began on a warm September day in 1884 when a group of sportsmen met in Philadelphia to form what became the American Kennel Club. They were following the lead of their British counterparts who, a decade earlier, had established the first all-breed kennel club, called, appropriately, the Kennel Club. The AKC soon established a registry that originally included 1,400 pedigree dogs representing eight breeds. As in England, the growth of enthusiasm for purebreds in the United States was spectacular. Dog shows began as a pastime of the landed gentry, but in the second half of the eighteenth century, the dog fancy caught the interest of the growing middle class. It was a classic case of fashion trickle-down from the rich to the wanna-be-rich.

Between 1900 and 1939, annual AKC registrations rose from 5,000 to 80,000. The big craze for purebreds came after World War II when the proportion of American dogs that were purebreds jumped from 5% to 50%, and registrations were growing fifteen times faster than the human population of the United States.

I trace the explosion in pedigree dogs, in part, to the post-World War II G.I. Bill. It enabled millions of Americans to buy homes in the suburbs with dog-friendly yards. My family was typical. My father, a bomber pilot, came home from the war in 1945 and got a job flying for an airline. With a low-interest loan from the Veterans Administration, my parents bought a house with a lawn and a beagle for my sister and me to play with. We were proud that our dog Bosco had "papers" from the AKC. In truth, no one in our family really knew what AKC registration meant, but it sounded good to us. Bosco was officially an aristocrat.

By 1970, the AKC was processing a million new registrations every year and was the world's largest registry of purebred dogs. The '80s and '90s were flush times for the organization. In 1990, half of the eligible dogs in the United States were registered with the AKC. The AKC outgrew their Manhattan headquarters, and most operations moved to a glass and steel building nestled in the woods of North Carolina's Research Triangle Park. In 2007, the AKC, a nonprofit organization, took in $72 million, about half of which was generated by puppy registration fees.

Dark clouds, however, loomed. Long controversial, the AKC took a major hit with the publication in 1990 of a devastating critique in the *Atlantic Monthly*. The author, Mark Derr, depicted the AKC as an elitist and secretive organization myopically focused on profits derived from the overbreeding of dogs for good looks. A few years later, registrations began a precipitous decline. Over the next three decades, AKC registrations plummeted 50% from their 1995 peak of a million and a half purebred puppies each year. In recent years, annual registrations seem to have stabilized at about 500,000 a year.

In addition to competition from other dog registries, the AKC's fall from grace was a response to changing public attitudes about the desirability of genetic purity in dogs. An AKC conformation show is a curious mix of eugenics and philosophy; the breeders are chasing an elusive Platonic ideal. Show dog people tell you that their sport is ultimately about improving the breed, bringing it closer and closer to perfection. Perfection in the show world is defined by a written code—the breed standard. According to the AKC standard, a Yorkie with a one-inch white spot on its chest is OK. A Yorkie with a two-inch white spot on its chest is disqualified. An akita must have black lips, a pink tongue, and ears that can touch its upper eye rim. One of my favorite rules dictates that a Clumber spaniel be "an intelligent and independent thinker."

All the breed standards pay lip service to a dog's temperament, but there is more emphasis on the color of the rims of its eyes and shape of its head than on traits that would make it fun to live with. I asked a handler at the Asheville show how much it would cost me to purchase a show-quality silky terrier. He said two to three thousand dollars, but then he added that he could put me in touch with a breeder who would sell me a "pet quality" silky for around $800. In the dog show world, "pet quality" means "loser."

But a good pet is exactly what 90% of puppy purchasers are looking for. In other words, the animals that breeders are trying to produce are different from the animals that most consumers want to buy. And, in their efforts to create the perfect dog, professional breeders have produced an animal whose elegant beauty is literally skin deep; a dog that looks terrific until you peer under the hood. That is the paradox of the show ring.

Mary Jean and I are suckers for big, good-natured dogs, dogs you can trust, dogs that like children. That's why we are drawn to Labs and golden retrievers. We have had three, and we have loved all of them. In each case, however, the hereditary burdens of their tribe came bundled with their registration papers. Both Molly and Tsali, our Labs, developed crippling hip dysplasia. In Tsali's case, it was so bad that we had her euthanized when getting up in the morning became an exercise in pain management. Dixie inherited a full complement of golden retriever maladies—dermatitis, hip problems, hypothyroidism (a diagnostic sign of which is "has a tragic look"), and congestive heart failure, the condition that finally did her in. Hereditary diseases are the rule, not the exception, among purebreds. Over 350 disorders lurk among the 19,000 genes that dogs carry around. Because many of these are shared with humans, purebred dogs have become a favorite animal model for the study of human diseases such as narcolepsy, epilepsy, and cancer.

Pedigree dogs are especially susceptible to genetic disorders for several reasons. Some are the consequence of intentional selection for physical deformities. Take the English bulldog. Beginning in the late nineteenth century, bulldog fanciers transformed an animal bred to latch on to the nose of a raging bull into a household pet. Bulldogs with monstrous brachycephalic heads and pushed-in faces became fashionable. This look was achieved by selecting for a skeletal malformation called *chondrodystrophy*. These distortions of the head and face produced a host of problems, including the inability to deliver puppies through the birth canal, labored breathing, snoring, and sleep apnea. Similar artificial selection for exaggerated structural features was responsible for bad hips in German shepherds (the consequence of breeding for sloping hindquarters) and bad backs in dachshunds. Indeed, a team of veterinary researchers found that eighty-four canine inherited disorders were directly or indirectly caused by breeding to standards dictated by kennel clubs.

Most genetic problems in pedigree dogs, however, are the by-product of inbreeding, not of selection for morphological weirdness. The majority of the 400 or so existing breeds of dogs were developed in the last 150 years from small pools of sires. All of the 20,000 AKC registered Portuguese water dogs trace their ancestry to thirty-one animals, and 90% of water dog genes come from only ten dogs. Inbreeding from a limited gene pool means that a puppy is more apt to inherit harmful recessive alleles from both its parents. Bad genes in just one ancestral line can do a lot of damage to a lot of dogs. This is known as the popular sire effect. Springer spaniels, for example, have the annoying habit of biting the hand that feeds them. Researchers at Cornell and the University of Pennsylvania have linked the occurrence of owner-directed aggression in this breed to one kennel, and more specifically, to a single dog. (The researchers also found that lines of springer spaniels bred for the show ring are more apt to turn on their owners than are springers bred for hunting.)

When 25% of a product is defective, you would expect a consumer revolt. As the bloom on registered purebreds has faded, the mixed-breed dog, which until the 1950s was the most popular pet in America, has regained its cultural panache. In the 1980s, animal protection groups began to promote the adoption of dogs relinquished to animal shelters, regardless of their pedigree, as morally superior to purchasing a purebred puppy. When in 2009, then Vice President Joe Biden's family picked a purebred German shepherd puppy to take with them to Washington, the headline on PETA's blog read "Joe Biden Buys One, Gets One Killed." The Bidens suddenly decided that they really needed a second dog, this time a rescue dog.

ARE WE RUNNING OUT OF GOOD DOGS?

As with other shifts in our tastes in dogs, the desirability of owning mixed-breed and rescue dogs has had an unanticipated consequence: As the number of people who want to adopt shelter dogs has gone up, the supply has gone down—way down. In 1970, twenty-three million dogs and cats were euthanized in animal shelters in the United States. By 2020, the number was less than two million. This drastic reduction in the number of abandoned pets is the result of the most successful campaign in the history of the modern animal protection movement: the spay and neuter crusade.

One result of our rush to pluck the gonads from every household pet is that America may be running out of dogs. Richard Avanzino, former president of Maddie's Fund, a foundation that aims to eliminate euthanasia in animal shelters, worries that the slack will be taken up by overcrowded puppy farms in China and Mexico. Others in the animal rescue community are not so sure that there is a dog shortage. They argue that we have plenty of adoptable dogs, they are just in the wrong parts of the country.

By some estimates, as many as 90% of adoptable dogs in some parts of the Northeast are ex-pats from the South. The abundance of unwanted dogs in southern states reflects regional differences in attitudes toward neutering pets. Spay and neuter campaigns have been much more successful in the Northeast, Midwest, and on the West Coast than in my part of the country. The per capita euthanasia rate of unwanted pets is forty times higher in North Carolina, for example, than it is in Connecticut. When the local animal rescue group in Jackson County, North Carolina, tried to prod the town council into enacting a mandatory spay and neuter law, it was a non-starter. People in the rural South don't like restrictions on their dogs any more than they like zoning or gun control.

My friend Jill's day job is teaching history to college freshmen, but her passion is finding homes for abandoned dogs. Over the years she has saved 2,000 dogs from the needle. Once a month, she plays God. It's heartbreaking. Jill walks the aisles of our local shelter carrying a fake arm made out of rubber. If she thinks a dog has a future, that someone could fall in love with it, she runs the dog through a standardized behavioral screening test. That's what the arm is for. She puts a bowl of food right in front of the dog's nose. Then, once it starts eating, she takes the food away with the rubber hand. If the dog growls or goes for the hand, it is toast.

Most dogs don't make the cut. Some of them bite the hand. Others are unadoptable because they are old or ugly. Small dogs are "in," and it has gotten harder to place big dogs, especially the black ones. Jill also avoids pit bulls. Too often, she says, they are adopted by the wrong kind of people. Almost all the animals she rejects will be euthanized. The dogs she deems adoptable will be vaccinated against rabies and parvo. Two weeks later, they will join two dozen other dogs in the back of a truck headed up I-95, riding the underground dog railway to suburbs in the Northeast.

Not all of Jill's migrants work out. She once got a call from an animal shelter near Greenwich, Connecticut, that was not happy with one of the dogs Jill had sent them, an Australian shepherd mix. He did OK on the rubber hand test, but he got seriously cranky once the truck passed the Mason-Dixon Line. The shelter gave her a choice: Either you come and get the dog, or we euthanize it. Jill got in her car that evening, drove fourteen hours up the interstate, picked up the dog, turned around, and drove fourteen hours home. It was a wasted effort. The Yankees were right; the dog was dangerous. A month later, Jill had to put him down.

THE HUMAN-DOG RELATIONSHIP IS A MIXED BAG

The decision of a handful of wolves between 15,000 and 45,000 years ago to entrust the destiny of their species to humans has been a mixed bag. Over time, the descendants of these self-tamed wolves weaseled their way into our hearts and the fabric of human society. The upside was a steady food supply and a warm fire to curl up by. The fact that the dogs' bipedal benefactors cooked a puppy now and then seemed a small price to pay for the security of home and hearth. But then came the leash, the collar, the fighting pit, obedience school, the animal shelter. Finally, there was the ignominy of genetic sculpting, in which humans transformed the basic canine form—a model of grace and efficiency—into shapes, shades, and sizes never found in nature. Over the last 150 years, we have jiggered the dog genome into a bewildering array of animals that look magnificent but which, in the end, are like modern tomatoes, a triumph of style over substance. We now choose our animal companions using the same facile consumer psychology that we apply to choosing the latest clothing styles and baby names. The transition from function to friend to fashion statement is complete.

At the same time, we love our dogs. We confide in them, buy them Christmas presents, take them on vacations, make websites for them, and treat them like our children. They sleep in our beds. We grieve when they die. The paradoxes of the human-dog relationship are particularly evident in the lives of people who are the most devoted to their animals. I admire the passion of purebred dog fanciers whose lives are focused on improving their breed. Their love for dogs is deep and real. But in their efforts to create the perfect dog, they have produced millions of animals with itchy skin, skulls that are too big, hearts that are too small, and hips that always hurt. Animals that suffer.

Nancy Brown's mission in life is to make her posse of wolf-dogs as happy as possible. But the hard truth is that these creatures are an uneasy mix of the wild and the tame, destined to spend their lives in the confines of a chain-link pen.

Jill knows two dogs will be euthanized in my county for every one that she finds a home for. She bears the moral burden of deciding which ones will live and which ones will die. Her commitment to dogs that no one else cares about has restricted her social life and interfered with her career. She looks sad most of the time. She got it right when she told me, "This is not something that you can do a little of."

My own relationship with dogs is also complicated. I have loved all the dogs we have lived with, even Maggie, a Benji look-alike we adopted from our local shelter. She was the smartest dog I ever met. Maggie's innocent face, however, belied a mean streak that got worse as she got older. She terrorized Tsali, our aging Lab, and eventually bit every member of the family, some several times. I consulted some of the best animal behaviorists in the country to no avail. One afternoon, Maggie savagely attacked a friend who was visiting. It was the last straw. Dr. Shields, our veterinarian, told me that I needed to face reality—it was time to put her down.

I remember that last long drive to the vet's office like it was yesterday. It's been ten years, but her death still weighs heavy on me. Maybe that's why we have a cat now. But I still miss having a dog around the house.

"Prom Queen Kills First Deer on Sixteenth Birthday"

GENDER AND THE HUMAN-ANIMAL RELATIONSHIP

As a society, we expect women to respond to the suffering of animals; we see that as a "natural" part of womanhood.

—BRIAN LUKE, PHILOSOPHER

Men enjoy hunting and killing, and these activities are continued in sports even when they are no longer economically necessary.

—SHERWOOD WASHBURN, ANTHROPOLOGIST

When I began my first forays into anthrozoology, I was struck by how differently men and women interacted with animals. For instance, I once interviewed veterinary school students about their reactions to moral issues such as euthanizing healthy animals. One of the women, Elizabeth, told me, "I cried the first time and I cried the fifteenth time. It hasn't gotten any easier over time, but I have learned to mask my feelings in front of the client, to be strong for them." Her male counterpart, William, had a completely different response. "Euthanasia doesn't affect me," he said. "I wonder about myself sometimes, but, honestly, I don't feel bothered by it."

After hearing a lot of similar statements, I came up with an obvious hypothesis; namely, women are nice to animals and men, well . . . not so much. Then I began to stumble on anomalies like Evelyn Clancy and Bill Gibson who don't fit our stereotypes about how the sexes relate to other species.

WHEN STEREOTYPES FAIL: STUDYING THE GENDER BENDERS

One afternoon, a graduate student named Amy Early walked into my office and told me she wanted to write her thesis on women hunters. It was good timing. I had just read a story in the *Sylva Herald and Ruralite*, the town newspaper, about a young woman who, on her sixteenth birthday, put a .30–30 slug in the chest of an eight-point buck. In addition to being an ace hunter, she was also the high school prom queen. I knew that 90% of hunters in the United States are men, but that the number of women who enjoy taking to the woods with guns is increasing. The newspaper story made me wonder how women who hunt negotiate the delicate line between nurturance—the traditional female role expectation—and their desire to kill animals.

My eyes lit up when Amy told me about her thesis idea. "Great project. Let's run with it."

One of the first people Amy interviewed was Evelyn Clancy, a middle-aged housewife who loved animals and also loved to hunt. Her passion was big game. She and her husband had been on several African safaris, but she had never bagged a zebra. One afternoon, Evelyn and her hunting guide Anthony came upon a large zebra herd. Whispering, she asked Anthony which of the animals she should shoot. He pointed out a big male. She snapped a bullet into the chamber, got the animal in the crosshairs of the scope, took a breath, held it, and squeezed the trigger. The zebra dropped.

Anthony turned and said, "That's not the right one." They walked over to the zebra, and Evelyn burst into tears. It was a female.

"I had shot a female," she told Amy. "I cried all evening, all night. I had tears in my eyes the next morning. It was the only time I ever shot the wrong animal. I felt awful about killing her. She may have been a mother. She may have had little colts running around. She could be pregnant. That's the reason it upset me so much, shooting the wrong zebra."

Evelyn's story says a lot about the complexity of gender in our interactions with other species. Her excitement at bagging her first zebra turned to despair the moment she realized she had killed a creature that she identified with. How many male hunters would cry all night because they had shot the wrong animal? And if they did, how many would admit it?

Bill Gibson represents a different type of gender bender. He is the skeptical, hard-nosed research psychologist who inhabits the office next to mine. One afternoon I asked him about the faded photograph of a dog he keeps on his desk. Bill is a math guy, a left-brainer. He designs computer systems that can look at your eyeballs and tell which part of your cortex is working overtime. But when he started talking about Blue, a shepherd-husky mix, he melted.

He told me that he named Blue because the dog had one blue eye and one brown eye. He bought Blue back in '84 from a guy who ran a puppy mill. For eleven years, Blue was Bill's best friend. "Blue would look me straight in the eye. He knew things," Bill said.

He told me how Blue had suffered through cancer of the bowel, and the prognosis was bleak.

"I was with him at the end," Bill said. "I put a nice blanket on him and gave him a kiss good-bye and talked to him. I had him cremated. I kept the ashes for a while and then later spread them on a mountain. I was a mess. Sometimes I would break into tears. It was embarrassing."

Bill started to choke up. "I still think about him every day. I loved him more than I've loved any person in the world. I'll probably think of him on my death bed."

Evelyn and Bill illustrate the difficulty of making simple generalizations about men, women, and animals. Cases like these convinced me that my initial hypothesis about sex roles and how we treat animals was wrong—clearly, some men are more sensitive to the plight of animals than some women. But then there are the obvious instances in which sex differences do fit gender stereotypes. For instance, all the cockfighters I know are men, and a large majority of animal rights activists are women. I decided to take a more systematic approach to investigating how gender affects our relationships with other species. I started amassing every study I could find on the topic. This quest took me, literally, from A (Animal Activism) to Z (Zoophilia), and I eventually published an article summarizing my findings. I discovered that some aspects of our relationships with animals are profoundly affected by gender. But I also found that both the direction of sex differences and their size depend on the type of relationship we are talking about.

WHICH SEX LOVES PETS THE MOST?

Surprisingly, gender differences in attachment to pets are smaller than most people think. In the United States, roughly equal numbers of men and women own companion animals and they are just as likely to buy holiday presents for their dogs and cats. The sexes are even similar in whether they let their animals sleep in their bed. (Women get the nod, but not by much.) But are men as bonded to their pets as women are?

Anthrozoologists have developed standardized questionnaires to assess how much individuals love their pets. For example, one widely used scale asks people how much they agree or disagree with statements such as "I would do almost anything to take care of my pet," and "My pet means more to me than any of my friends." Most studies have found that when it comes to pet love, women seem to have an edge over men, but the difference is surprisingly small. I examined a dozen studies that reported sex differences in attachment to pets. True, women scored higher than men in eight of them, but the difference between the attachment score of the average man and the average woman was usually negligible. And this finding also applies to children.

In some aspects of caring and interacting with pets, however, gender differences do show up. Women, for instance, are more disposed than men to use baby talk when interacting with their dogs. You will not be surprised to learn that women are twice as likely as men to dress their pets in little outfits. And, as is the case in many spheres of human life, women do more than their fair share of pet-related chores. In three out of four American homes, women are the ones who usually feed the family dog and clean out the cat's litter box. Women also make up about 85% of veterinarians' clients. (Several veterinarians told me that when men bring pets into their clinics, they often carry notes from their wives with them spelling out exactly what is ailing Spot or Fluffy.)

I have always assumed that males and females play with pets differently. After all, boys are more apt to engage in the mock wrestling and hand-to-hand combat that developmental psychologists call *rough-and-tumble play*. In my house, there was a gender bias on how we relate to our family pets. For example, Mary Jean and our twin daughters, Betsy and Katie, would gently pat our dogs on the head while Adam and I were more apt to wrestle and play chase and pull-and-tug with them.

There is, however, little evidence that my hypothesis applies to anyone outside the Herzog household. Italian researchers found that men and women were not any different in how they played with their dogs. Similarly, Austrian researchers did not find any sex differences in how owners interacted with their dogs in a variety of situations. Gail Melson, a developmental psychologist, asked parents to estimate how much interest their children had in playing with younger children, babies, stuffed animals, baby dolls, and pets. As you would expect, girls were much more interested than boys were in babies, dolls, and stuffed animals. But she found no sex differences in how kids played with or nurtured their pets. Gail argued that for boys, pets are often the only vehicles that give them experience in caring for another living being.

Women, however, are more susceptible than men to cute creatures. British researchers reported that two groups of women are particularly sensitive to differences in the cuteness of human infants: those of reproductive age and those taking birth control pills that artificially raised their levels of the hormones progesterone and estrogen. In a field study, University of California at Santa Barbara researchers examined changes in the attractiveness of a golden retriever puppy named Goldie as she matured. For five months, they took Goldie to a busy spot on campus where she would sit for an hour with her "owner" (actually an assistant in the study), while the researchers tallied the number of passersby who came over to pet or play with her. Goldie's ability to seduce strangers decreased precipitously as she morphed from puppy to adult dog. Her drop in popularity was especially steep among women. When Goldie was at her cutest, women were twice as likely as men to chat her up. But by the end of the study, the number of women who stopped by to stroke her head and say hi had dropped 95%, and the sex difference had completely disappeared.

SEX DIFFERENCES IN ATTITUDES TOWARD ANIMALS

Stephen Kellert of Yale University spent his career investigating the peculiarities of human attitudes toward other species. He consistently found that women are more concerned with protecting animals than men are. At the same time, phobias of creatures like snakes and spiders are three times more common in women than men. And while men know more about animals than women, they tend to appreciate animals for what Kellert calls "practical and recreational reasons." In other words, men are more likely to approve of killing animals for fun and profit.

Building on Kellert's studies, my colleagues Robbie Pittman and Nancy Betchart and I developed a questionnaire we called the Animal Attitude Scale to examine differences in attitudes toward animals. Like Kellert, we found that women were more concerned about animal welfare than men. In addition, men who had a more feminine *sex-role orientation* tended to care more about the treatment of other species. And, as you might expect, women reported they would be less comfortable than men touching creepy crawlies like worms, snakes, and spiders. Many studies have now reported similar sex differences in attitudes toward and beliefs about other species.

Sex differences in attitudes toward animals exist across cultures. Linda Pifer of the Chicago Academy of Sciences asked adults in the United States, Japan, and thirteen European countries how they felt about using dogs and chimpanzees in experiments aimed at developing treatments for human afflictions. In every country, women were more opposed to animal research than men. Swedish researchers reviewed dozens of studies conducted worldwide and did not find a single country in which more women supported animal research than men.

Glib generalizations about sex differences and animal attitudes, however, can be misleading. The fact is that in public attitudes about the use of animals, men and women are more similar than they are different. The National Opinion Research Center asked a large random sample of men and women how they felt about this statement: *It is right to use animals for medical testing if it might save human lives.* They found that more men than women "strongly agreed" with the statement and more women "strongly disagreed." Most people, however, were in the middle on this issue, and some women were more supportive of animal experimentation than some of the men. More recently, researchers found that men were more likely than women to support the use of each of ten species of animals in research. The differences between the sexes, however, were small. Men and women don't differ much in the degree to which they attribute mental states to other species either. Together, these findings illustrate a fundamental attribute of human gender differences—in nearly every human psychological characteristic, men and women overlap. In short, the individual differences within the sexes are greater than the differences between the sexes.

MORE WOMEN THAN MEN DO THINGS FOR ANIMALS

While men and women are similar when it comes to *attitudes* about animal welfare, the sexes are not alike when it comes to doing things on behalf of other species.

Take, for example, Leslie Irvine, a University of Colorado sociologist. A few days after Hurricane Katrina leveled the coast of Louisiana and Mississippi, Leslie and three of her friends hopped on a plane. Their destination was the Lamar-Dixon Expo Center in Gonzales, the main staging area for animal rescue operations around New Orleans. The Colorado team was unprepared for the chaos they encountered. That night, Leslie wrote in her field notes, "Who can imagine the sound of a thousand dogs barking? Until today, the question would have seemed like a perverse *koan*. But now that I know what a thousand dogs sound like, I wish everyone could hear. It sounds like futility, helplessness, and the desperation of this undertaking. . . . I am sure it will haunt me for a very long time." Six days later, Leslie became a Katrina victim herself. Mentally and physically exhausted by nearly a week of animal rescue, she collapsed and was taken to the hospital. Years later, she was still haunted by the sound of a thousand dogs barking. There were lots of heroes like Leslie who saved the lost animals of Katrina, and 80% of them were women.

Women are much more likely than men to change their lives because they care deeply about animal suffering. The modern animal protection movement has recruited women in droves; every study of the sociology of animal liberation has found that three to four times as many women as men boycott circuses, march against animal experimentation, and drive cars with MEAT IS MURDER bumper stickers. The sex ratio of individuals involved in animal protection has not changed at all over the last 150 years. Even in the Victorian era, four out of five members of animal welfare organizations in the United States and Great Britain were female.

Indeed, women dominate nearly every aspect of grassroots animal protection. They make up 85% of the membership of the two largest mainstream animal protectionist organizations in the United States, the ASPCA and the Humane Society of the United States. A study of dog rescuers found that women outnumbered men eleven to one. And three times more female high school students than males called the National Anti-Vivisection Society's Dissection Hotline because they wanted to opt out of biology dissection labs. More women than men give up eating animals, and they are more likely to stick to their vegetarian or vegan diets.

At the National Zoo in Washington, D.C., hundreds of dedicated volunteers, most of them women, stand next to exhibits for hours on end, patiently explaining a hundred times a day that the alligator snapping turtle really is alive even though it looks like a hundred-pound hunk of moss-covered granite, that Spike the elephant is forty-one years old, and that a pygmy hippo is not the same as a baby hippo. You see even bigger sex differences among the keepers. As I strolled around the primate facility one snowy January day, I ran into a zoo employee. She told me that every one of the eleven keepers who cared for the chimpanzees, gorillas, and orangutans in the Great Ape House were female.

Like primate keepers at the National Zoo, women are attracted to many professions involving the care and treatment of animals. Women certainly predominate researchers who study human-animal interactions. Eighty percent of the members of the International Society for Anthrozoology are women, as were 80% of the authors of articles published in the society's journal *Anthrozoos* in 2019. And, in 2021, thirteen of the fifteen members of the society's board of directors were women, including the president, the president-elect, and the past president. Women will likely continue to dominate the field of human-animal studies. Nearly 90% of the 200 graduates of the master's degree program in Animals and Public Policy at Tufts University have been women.

Veterinary medicine illustrates the rapid changes in the roles of women in American society and gender differences in concern for animals. In 1960, veterinary medicine was essentially an all-male profession—fewer than 2% of vets in the United States were women. This skewed sex ratio began to reverse in the 1970s. By the time I was studying how veterinary students dealt with moral conflicts in the mid-1980s, half of the students at the University of Tennessee vet school were women. Today, 60% of veterinarians in the United States and 80% of vet students are women.

Why have women in droves flocked to fields like veterinary medicine, animal shelter work, professional dog training, and even laboratory animal care? I ran this question by Diana Fleischman, an evolutionary psychologist who studies animal issues. She believes these professions align with a natural feminine propensity to care for animals. She emailed me:

> From my perspective, the culture has changed in ways that women were more likely to go into the workforce than to stay at home and be homemakers. And they chose careers that were interesting to them. Animal issues have very frequently been entangled with feminist politics because women have an interest in both areas. As women came to be equally or over-represented in fields like biology, psychology, and sociology starting in the 80s, they also came to be overrepresented in animal issues.

Makes sense to me.

THE DARK SIDE OF THE BOND: SEX DIFFERENCES IN ANIMAL CRUELTY

But don't make the mistake of thinking that all women are nice to animals.

Take the case of Joanne Hinojosa who, on a Thursday morning, was arguing with her estranged husband in front of his home in South Austin, Texas. At some point during their spat, things took a turn for the worse. When Joanne took a swing at him, he took off down the street to call the cops. That's when Joanne went after his dog. She carried Marti, his twenty-pound mixed-breed female, into the kitchen and started stabbing her. When the police arrived, they found Marti lying in a pool of blood with a knife sticking out of her left side. The dog had been stabbed twenty-seven times. They rushed her to the Ben White Pet Hospital for emergency surgery, but it was too late. Hinojosa's lawyer claimed his client suffered from post-traumatic stress disorder. She pled guilty to animal cruelty and was sentenced to six months in jail and anger management training.

This was not an isolated event. Pets are often caught up in domestic violence disputes. Developmental psychologist Frank Ascione found that over 70% of battered women he studied said that their partners had abused, threatened to abuse, or killed a family pet. Their violence ranged from shooting the family dog to setting the children's kitten on fire. What is unusual about the Hinojosa case is that the perpetrator was a woman.

Data compiled by Pet-abuse.com, a website that tracked media reports of animal cruelty cases, offered a window into sex differences in animal cruelty. The psychologist Kathy Garbasi reported that men were perpetrators in 70% of the 15,000 cases in the database. But this figure is misleading. The sexes were fairly similar in the frequency with which they were charged with animal neglect—which is usually the result of apathy, poor judgment, or stupidity rather than malicious cruelty. Men commit nearly all of the really nasty offenses against animals: 84% of stranglings; 91% of burnings; 92% of stompings to death; and 94% each of beatings, hangings, and shootings.

PEOPLE WHO LOVE ANIMALS TOO MUCH: GENDER AND ANIMAL HOARDING

There is an exception to the rule that most animal abusers are men. Three times as many women as men get caught up in animal hoarding. I once invited a woman who was an ordained minister to be a guest speaker in a class I was teaching. After her lecture, she mentioned to me that her house was for sale. I perked up. Mary Jean and I were in the market for a new home. I drove out to look at the house the next day. It was halfway up the side of a mountain near Greens Creek. The property had privacy, a southern exposure, and a big view toward north Georgia. Just what we were looking for. I had my hopes up. Then I opened the front door.

"Overpowering" does not quite capture the stench of feces and urine in the house. Cats and clothing were strewn about the living room. A forty-pound bag of dried cat chow lay spilled open in the middle of the floor. There was no place to sit. The woman, who had seemed perfectly normal when I talked to her the day before, was in a tizzy, manically apologizing to me that things had gotten a bit out of hand, that she had tried to straighten things up for my visit. I counted two dozen cats but there were probably others roaming the nearby woods. I did not stay long, and we did not buy the house. We probably could have gotten it cheap.

An interdisciplinary group of experts estimated that about 5,000 cases of hoarding involving 250,000 animals are reported in the United States each year. People usually don't think that hoarding is as cruel as beating and shooting animals, but from the animals' perspective, it can be worse because the suffering can go on for years.

The stereotyped animal hoarder is a crazy old lady living alone whose neighbors call the health department when they finally tire of the noise and the stink. Is this image accurate? Generally, yes. Seventy-five to eighty-five percent of hoarders are women. Most of them live alone in squalor, and half are over sixty. Cats are more commonly hoarded than dogs, followed by birds and small mammals. In about half the cases, animal hoarders also hoard objects. Hoarders nearly always seem oblivious to the horrid conditions they and their animals are living in.

Some animal hoarders are nuttier than others. According to Arnie Arluke and Celeste Killeen, authors of the book *Inside Hoarding: The Case of Barbara Erickson and Her 552 Dogs*, the woman I met would probably fit into the *incipient hoarder* category. With twenty or thirty animals living in her house, her life was just starting to fall apart. The numbers of animals confiscated from a hoarder's home can be unimaginable. The record is "The Great Bunny Rescue of 2006" in which nearly 1,700 rabbits were rescued from the backyard of a woman named Jackie Decker who lived in a two-bedroom house near Reno, Nevada.

A study on the public health implications of animal hoarding found that nearly all of the hoarders who had over one hundred animals in their homes were women. The living conditions of these hoarders ranged from lousy to horrifying. Things were particularly bad among those who lived alone. Over half of their houses lacked stoves, hot running water, or working sinks and toilets. Forty percent of the homes had no heat, and 80% did not contain a functional shower or refrigerator. The conditions for the animals living in these circumstances were dire—cats, dogs, pot-bellied pigs, rabbits, all emaciated and starving and ridden with parasites and disease, all running amok. First responders called in to clean up hoarding situations often encounter the half-eaten corpses of animals lying about.

Clinicians have come up with several explanations of why people hoard animals. The wildest theory is that cat hoarding can be caused by toxoplasmosis infection. The idea that hoarding is the result of a para-site that rewires your brain is intriguing. Tox-infected rodents suddenly become attracted to cats and can't seem to get enough of the smell of cat urine. In humans, tox infection has been associated with mental ill-ness and neuroticism. It is not a far stretch to imagine that the parasite might cloud the judgment of people with lots of cats and perhaps even zap out enough neurons that they become inured to the smell of tomcat pee, an odor most people find intensely aversive. Unfortunately, there is little evidence to support this theory. And one study, surprisingly, found that dog hoarders had lower rates of exposure to tox than non-hoarders in the same communities. Conventional theories link hoarding to dementia, obsessive-compulsive disorder, addictive personality, defects in social at-tachments, or delusional thinking. Indeed, hoarders are usually convinced that their animals are happy and that the hoarder has a special ability to communicate with them.

While anthrozoologists don't know the causes of hoarding, research-ers agree that it is nearly impossible to cure. The recidivism rate among hoarders is high. Most hoarders are reluctant to get therapy and resistant to change. In the Reno bunny case, for instance, animal control officers had confiscated 500 rabbits from the same property just four years earlier. Though most of the animals had to be euthanized, a local judge treated the case as a joke—he refused to issue a court order that would have kept Jackie Decker from obtaining even more animals.

I suspect most hoarding results from a perfect storm of the well-intended desire to save animals and the inability to draw lines in the moral sand. This means that individuals who are drawn to animal rescue are at special risk. Indeed, one of the subtypes of hoarding is called "rescue hoarding." These cases often involve extremely large numbers of animals. An article in the *Veterinary Journal* reported four cases of hoarders who ran illegal cat "sanctuaries" containing between 387 and 697 cats.

For a couple of years, my students and I interviewed members of animal rescue organizations. Most of them were perfectly normal, and some were saints, but occasionally we ran into red flags. During the interviews, we asked rescuers how many pets they had. Most people said something like, "two dogs, a cat, and a parrot." But sometimes when we ask about their pets, a rescuer would look down, laugh a little, and say something like, "Oh, hmm . . . too many, I guess." These were the people who did not want us to interview them in their homes.

I got a twinkling of insight into how a person might not be able to just say no, how they might *have* to bring just one more animal into their house when I was given a behind-the-scenes tour of a municipal animal shelter. Immediately, I was drawn to a young boxer lying on his side, panting, in a cage in the quarantine room. He lifted his head, and looked at me with the world's saddest eyes that said *Take me out of here, please?* I could feel my chest getting tight, and I was a goner. If he had not been sick and in quarantine, I would have brought that dog home on the spot.

Eight thousand dogs and cats entered that shelter every year. Forty percent of them would leave with a new owner wearing a big grin. The other 60%, including most of the cats, and nearly all the large black dogs and pit bull mixes, would get a couple of cc's of sodium pentobarbital injected into the cephalic vein of their foreleg and slip away in a matter of seconds.

My tour guide Becky, the director of the facility, had worked in shelters for fifteen years and she loved animals. She really did. As we squeezed past rows of stainless steel cages, she pointed to a Treeing Walker, a type of coon hound, and told me that it was found running loose near Big Oak

Gap, and then she called out an orange tabby cat by its name. The back rooms were crowded and my ears were ringing from the incessant barking of the dogs. It was too loud to talk. Some of the animals looked fine, others were terrified.

Becky's passion was making life better for unwanted animals. She told me about her own pets—three cats, three dogs, and three birds. The paradox of her profession became apparent when I asked, "How many dogs and cats have you euthanized over the years?"

She looked at me like I was an idiot.

"Over a thousand?" I asked meekly.

After a pause, she said, "At least."

"How do you stay sane?"

"Somebody has to do it. I don't obsess about it," she said.

Then she showed me a message that had appeared on an Internet LISTSERV a couple of days previously. It was written by a shelter director who went home every night and cried. She was bitter. "I hate my job," she wrote. "I hate that it exists and I hate that it will always be there unless you people make some changes and realize that the lives you are affecting go much further than the pets you dump at a shelter. I do my best to save every life I can but there are more animals coming in than there are homes."

Becky told me that the other director is in the wrong line of work.

Remarkably, Becky remained cheerful and passionate about her career. She was the right person for the job. But this was not true of all volunteers she supervised, 85% of whom were women. She said she had to keep her eye on a few of them, the ones that had hoarder potential, the ones that didn't have the moral strength to do the work.

THE NATURE (AND NURTURE) OF SEX DIFFERENCES IN OUR INTERACTIONS WITH ANIMALS

After reading hundreds of articles documenting sex differences and similarities in our relationships with animals, I have come to several conclusions.

First, women tend to have more of a soft spot for animals than men. No surprise there. Second, some of the stereotypes about the size of the differences between men and women are wrong. There is not much difference at all between the sexes in the frequency that they live with pets and in the way boys and girls play with dogs and cats. Sex differences in attitudes toward issues like animal research are a bit larger, but there is a lot of overlap between the sexes. The big sex differences emerge when you look at the extremes—for instance, animal protectionists and animal abusers.

Everyone wants to know whether human sex differences are a result of nature or nurture. This is a loaded question and an intellectual loser. To think that complicated behaviors like the way we feel about animals are the result of either nature OR nurture exemplifies the myth of single causation. For example, both biology and culture influence sex differences in ethical judgments. A study involving over 300,000 people in 67 countries found that in all cultures, women scored higher than men on the values of care, fairness, and purity. But there were large cultural differences in the degree to which the sexes valued loyalty and obedience to authority.

Some animal activists argue that women are more involved than men in animal rights because both women and animals are victims of oppression. Certainly, some of the differences between men and women are rooted in exploitation, but biology also plays a role. Take hunting. In our closest living ancestor, chimpanzees, males hunt more than females. And hunting is defined as a male activity in every human culture. Well, nearly every human culture. In the Ituri Forest of the Democratic Republic of the Congo, Bambuti pygmy women help drive game into nets, and Matsés women in the Amazon Basin often accompany their husbands on hunting forays. They spot game and kill animals with spears and machetes. Because hunting trips offer privacy, a round of hot jungle sex can help the time go by when prey is scarce. For this reason, Matsés men, who are polygamous, take only one of their wives on hunting expeditions, and women complain when they don't get invited to "go hunting" enough. Cultures like these, however, are exceedingly rare. In nearly all human societies, hunting expeditions are as sexually integrated as the Green Bay Packers locker room during half-time.

Developmental psychologists have found that some sex differences show up so early that they are unlikely to be the result of socialization. Among kindergarten kids, boys show marked preferences for creatures like sharks, beetles, and alligators while girls are more attracted to butterflies and birds. Even in infancy, girls are better at discerning emotional expressions than boys. And (I know this is hard to believe) several studies have found that male and female monkeys show the same differences in preferences in toys as human children. Male monkeys are attracted to "boy" toys (trucks, for example) and female monkeys like to play with soft, cuddly objects. When sex differences in our perceptions and interactions with animals begin to show up is unclear. The developmental psychologist Vanessa LoBue tells me she has not found much evidence for sex differences in the responses of infants and young children to snakes and spiders.

Our body chemistry also affects our interaction with other species. Hormones affect empathy, a trait that affects how we think about and behave toward people and animals. One of these is oxytocin, a chemical that switches on maternal instincts and facilitates social bonding. In humans, oxytocin levels rise during pregnancy and spike during childbirth, breastfeeding, and sexual orgasm. Oxytocin is also related to sex differences in empathy. Men, for example, are not as good as women at reading the emotions hidden in a face, but some studies have found that a whiff of oxytocin can hone their emotional intelligence and make men more generous.

Is oxytocin the glue that cements the human-animal bond? Meg Daley Olmert thinks so. In her book, *Made for Each Other: The Biology of the Human-Animal Bond,* she writes that pets are "fountains of oxytocin" and that pet lovers go through their day with an "oxytocin glow." Her claim, however, was based on a single study that had only eighteen subjects. True, the researchers did find that oxytocin levels rose after people interacted with dogs—but they also found that oxytocin increased nearly as much when the subjects just sat quietly and read a book. Recent experiments on the role of oxytocin in our relationships with pets have reported mixed results. In one study, levels of the hormone increased in women who petted dogs, but went down in men. A group of Japanese researchers found that whether or not oxytocin increased when owners played with their dogs depended on how much the dog gazed at its owner during the interactions, as well as the owner's age. A recent study by Australian researchers reported, to the investigators' surprise, that levels of oxytocin in dogs did not change when they interacted with their owners. Further, oxytocin levels in owners were unrelated to how attached they were to their pets. The anthrozoologist John Bradshaw is an oxytocin skeptic. He thinks all the attention the hormone has gotten is "mostly hoopla." Bradshaw suggests endorphins, another family of feel-good hormones, are more important in generating the little rush I get when Tilly jumps into my lap and looks at me, telling me to scratch behind her ears.

DOES THE BELL CURVE EXPLAIN SEX DIFFERENCES IN OUR RELATIONSHIPS WITH ANIMALS?

The bottom line is that sex differences in our interactions with other species are the result of an inextricable mix of political, cultural, evolutionary, and even biochemical forces. Is there a way to make sense of both the large and small differences we see in how men and women think about and behave toward animals without resorting to clichés in the debate over nature and nurture?

Some years ago, I came across a little-known article in *The New Yorker* by *Tipping Point* author Malcolm Gladwell. The title was "The Sports Taboo: Why Blacks Are Like Boys and Whites Are Like Girls." In it, Gladwell argued that to get a handle on racial and sex differences you need to understand an important property of the bell-shaped curves statisticians call "normal distributions." Bell curves describe many psychological and biological phenomena. The basic idea is simple. For traits ranging from the scores on personality tests to the size of goldfinch beaks, most cases will fall near the middle of the pack with the numbers tapering off as you move toward the extremes of the distribution. IQ test scores are a good example of a bell-curve trait. The average IQ score in the United States is 100. While 50% of people have IQs greater than 100, only 2% score over 130 on IQ tests, and one person in 1,000 gets above 145.

Bell-curve thinking is sometimes—and wrongly—thought of as racist. That's because in a 1994 book called simply *The Bell Curve*, Richard Herrnstein, a psychologist, and Charles Murray, a political scientist, used normal distributions to support their contention that racial differences in IQ are inherited. But bell curves are just shapes. They say nothing about whether differences between groups are due to genes or environment.

While bell curves do not explain the ultimate causes of sex differences, they can help us understand why most animal activists are women and most animal abusers are men. My position is that many human sex differences, including how we treat animals, are simply the consequence of an elegant statistical principle that even most psychologists don't get. It is this: *When two bell curves overlap, even a small difference between the average scores of the groups will produce big differences at the extremes.*

Height is a good example. In the United States, the average man is 8% taller than the average woman. This does not sound like much, but the sex ratio gets more and more out of whack as we go toward the high and the low ends. For example, among people over five foot ten inches, there are thirty men for every woman, but the sex ratio shoots up to 2,000 to 1 when we look at people over six feet.

The statistical principle that small differences between the average male and female make for big sex differences at the extremes explains sex differences in many areas of human behavior. The fact that women are ten times as likely as men to have an eating disorder follows directly from the fact that the average American woman is a bit more concerned with body image than the average man. And the enormous sex difference in homicide rates is a consequence of the real but surprisingly small difference in the aggressive tendencies of the average man and woman.

Here is how the Herzog-stolen-from-Malcolm-Gladwell theory of sex differences plays out in human-animal relationships. Assume for a moment that Americans vary in a hypothetical psychological trait called "liking pets," and that the distribution of this trait is bell-shaped—most people are in the middle, but a small proportion of people pathologically dote on their animals and a few people truly hate animals. Assume also that the average woman scores slightly higher than the average man on this trait but—as is nearly always the case—there is a lot of overlap between the sexes. If my bell-curve theory is correct, as we move toward the pro-pet and anti-pet extremes, bigger and bigger gender differences should emerge.

This is exactly what we find. Among extreme pet lovers, say off-the-chart hoarders, women outnumber men ten to one, and among sadistic animal abusers the male-to-female ratio is even more skewed. Overlapping bell curves also explain why so many animal rights activists are women. Surveys of the general public show that women are more concerned with the welfare of animals than men are. The difference, however, is not all that big, and the magnitude of differences in attitudes toward animals within the sexes is a lot bigger than the differences between the average male and the average female. But, once again, as we move toward tails, the sexes go their different ways. On the pro-animal side, four times as many women as men donate money to the ASPCA, boycott circuses, and show up at animal rights demonstrations. On the anti-animal side, many more men than women find pleasure in shooting animals for sport.

The bell curve explains a wide array of sex differences in human-animal interactions. And it works regardless of whether evolution or culture is responsible for the inner voices that whisper "I feel so sorry for that cute little monkey in the cage" to a six-year-old girl visiting the zoo for the first time or "*Exhale slowly. Hold. Pull the trigger . . . NOW!*" to a teenage boy on his first father-son deer hunting expedition.

In the Eyes of the Beholder

THE COMPARATIVE CRUELTY OF HAPPY MEALS AND COCKFIGHTS

> The people who set one animal against another haven't got the guts to be bullies themselves. They're just secondhand cowards.
>
> —CLEVELAND AMORY, AUTHOR

> Cockfighting is the most humane, perhaps the only humane, sport there is.
>
> —CAPTAIN L. FITZ-BARNARD, COCKFIGHTER

I am driving toward Knoxville on Interstate 40 to interview Eddy Buckner, an old cockfighter I hung out with years ago when I was writing my doctoral dissertation on the behavior of chickens and the psychology of chicken fighters. Ten miles from the Tennessee line, I notice white feathers whizzing past my windshield. I speed up, pass a couple of eighteen wheelers and find myself behind two flatbed trucks, each loaded with thirty-four wire crates packed with live chickens headed for the slaughterhouse. The crates are about three feet wide, four feet long, and ten inches high, and look like they hold thirty or forty chickens each. I do the math in my head: At three birds per square foot of cage space, over 1,000 animals on each truck, the chickens crammed together like anchovies packed in a Trader Joe's jar of olive oil. It is fifty-five degrees outside, and the chickens are exposed to the wind and highway noise. The trucks are pushing eighty as we cross the state line into the aptly named Cocke County. Feathers flying, chickens shivering, hiding their heads under their wings, terrified. I think to myself, *Why don't federal animal welfare laws cover the interstate transport of commercial poultry?*

I tail the trucks for twenty miles, past the Wilton Springs exit, only a stone's throw from the abandoned Four-Forty Cockpit, which closed down in 2005 when the FBI put the hammer down on east Tennessee rooster fighters. It strikes me that while a lot of chickens have been cut down in the Four-Forty main pit over the last thirty years, a thousand times more have died to feed the tourists who will stop at the McDonald's just down the road for a six-piece McNugget Happy Meal. And I realize that the comparative ethics of fighting chickens versus eating them is more complicated than most people realize.

■ ■ ■

If Molly, our Labrador retriever, had not developed a taste for raw eggs, I would never have discovered the subterranean world of illicit cockfighting in my little community. Shortly after we moved to the mountains, Molly turned thief and took to sneaking into Hobart Presswood's henhouse, stealing eggs. I got an inkling that something was amiss when I drove home from work one afternoon and found Molly lying on the porch with an injured leg and egg, literally, on her face. The next day Mary Jean ran into Hobart's wife, Laney, in the grocery store and mentioned that Molly had been hurt but we did not know what happened to her. "Oh," Laney said. "Hobart caught her in our hen house again and shot her in the rear end. Don't worry none. He just peppered her with birdshot."

I was not mad at Hobart. After all, my dog was stealing his eggs. But I knew I had to do something fast. The solution, I figured, was to get my own chickens and teach Molly not to mess with them. I looked in the classifieds under "poultry" and found what I was looking for: *Chickens—$2.50 each. You catch them. Call R. L. Holcombe, Stony Fork*. The price was right and

Stony Fork was only a couple of miles away. I hopped in my truck and drove over the mountain. A dozen chickens were running loose in Mr. Holcomb's place including some small mousy brown hens and a couple of magnificent roosters. Mr. Holcomb told me they were gamecocks, that he had once been a serious cockfighter but now he only kept a few birds around for pets. After chasing them around his yard for an hour, I managed to catch a rooster and a couple of hens, but more importantly, I got him talking about chicken fighting. That's when I became aware that I was living amidst a world that was invisible to my fellow college professors.

Mr. Holcombe said that if I wanted to learn about game chickens, I needed to talk to Fabe Webb, a legend among western North Carolina cockfighters. The next morning, I gave him a call, and to my surprise, he said to come on over. A big man in his seventies with red hair and a ruddy face, Fabe lived half a mile up a rough dirt drive in a small white house surrounded on three sides by Pisgah National Forest. You could not see his place from the road, but you could hear the birds crowing a long way off. He had dozens of gamecocks, resplendent animals—some deep maroon, almost purple, with iridescent green hackles, some pure white, others black with orange necks—all of them individually housed in wire cages.

Fabe loved to talk chickens, and I began to drop by regularly. He would take me through his chicken yard, explaining the bloodlines of each rooster, tossing a bit of grain here and there, clucking to his birds. Now and then, he would point out a battle-scarred old gamecock, perhaps blind in one eye, and Fabe would puff up a bit, like one of his roosters, and say something like, "Now that one's a six-time winner." The truth was that Fabe had given up cockfighting years before I met him. But he still went to an occasional derby, sat in the bleachers, and bet on a couple of matches.

I was completely baffled. How could a person who loved birds so much participate in a brutal bloodsport that always ends in death? I brought up this contradiction one afternoon while we were sitting in his kitchen sipping some of western North Carolina's finest moonshine. Fabe offered to take me to a fight. He said he wanted to show me that cockfights were not the blood baths that city boys like me thought they were. It never happened. I kept putting him off, until one day, I read in the morning paper that Fabe had died. I think it was a heart attack.

Shortly after Fabe's death, the dean of the college that had hired me as a temporary instructor in the psychology department called me into his office. The good news was that he offered me a permanent job. But there was a catch—I would need to finish my PhD. All of a sudden, watching the chickens that I bought from Mr. Holcombe went from a casual pastime to my ticket to the tenure track. After reading up on poultry behavior, I convinced my advisor at the University of Tennessee to let me write my dissertation on chickens. I wanted to investigate breed differences in the temperaments of baby chicks, including gamefowl chicks. Two months later, I was running a minihatchery in my basement. I did not have any trouble getting fertile Rhode Island Red and White Leghorn eggs from the university's poultry science department, but getting my hands on purebred gamecock eggs was another matter. After contacting friends of friends of friends, I located a couple of Tennessee rooster fighters who were happy to oblige. And one of them, a man named Jim, invited me to tag along to an upcoming derby over in North Carolina. This time, I went.

A FIVE-COCK DERBY IN MADISON COUNTY

The fight was at the pit up the road from the abandoned Ebbs Chapel School in Madison County; from the outside, it looked like a big barn. We paid the admission fee, and Jim went over to talk to some of his buddies while I tried to make myself invisible. The air was hazy with cigarette smoke and heavy with the smell of coffee and hamburger patties frying on the grill in the refreshment stand. A hundred and fifty or so people were sitting in the bleachers or milling around. Among the spectators was a man in a wheelchair sitting next to his wife and son, who looked to be about twelve years old. Above the general din of conversation, the voice of the pit owner boomed over the PA system, telling the handlers awaiting the next match to bring their roosters to the main pit.

Each of the handlers was carrying a rooster that looked odd. That's because their combs and wattles had been cut off ("dubbed") and the feathers on their backs and sides trimmed so the birds would not overheat during the fight. One rooster was deep red. His opponent was black with pale yellow hackles. Like high school wrestlers, the cocks were matched for weight. Gaffs, the curved pointed steel blades that make cockfights deadly, were attached to the stubs of the natural spurs on the legs of the chickens with leather and waxed string. As was customary in Appalachia back then, they were fighting long heels—two and a half inch gaffs—rather than the shorter gaffs favored in the north or the slasher blades worn only on the left foot that Filipino and Hispanic cockfighters prefer.

A bald guy in the bleachers yelled out to no one in particular, "I'll lay a 25-to-20 on the Gray." A younger man sitting across the pit pointed to him and said, "You're on." Jim nodded toward a handful of men hanging out quietly together in a corner. He told me they were the high rollers that made big bets among themselves.

The referee was a fifty-year-old black man they called Doc, who by day, was a school janitor. He gave the signal, "Bill 'em up." The handlers cradled the birds in their arms and brought them together in the center of the pit, a circular arena about fifteen feet in diameter surrounded by a three-foot wire fence. When the cocks saw each other up close, the adrenaline kicked in, their hackles flared, and they went for one anothers' eyes. After the birds pecked at each other's heads for a few seconds, the handlers separated them and retreated behind the long score lines drawn in the dirt. They squatted down on their haunches, holding their birds, waiting. Doc yelled, "Pit 'em!" They released the roosters, and the fight began. All I saw was a blur of feathers.

Ten minutes later, the red cock's handler tossed his rooster's limp body into a barrel of dead chickens. By the time the bets were paid off, two more roosters were ready, and I heard Doc say, "Pit 'em!"

■ ■ ■

I dragged home at three in the morning, but I tossed and turned all night. At some unconscious level, I was trying to figure out what it all meant. Bob Dylan has a line that goes, "You know something is happening here but you don't know what it is—do you, Mr. Jones." That's how I felt.

The next morning over coffee, I told Mary Jean that I was going to shift gears and become an ethnographer for a while. I was certainly not the first researcher who tried to make sense of rooster fighting. Most of the others have been anthropologists looking for underlying meaning— totems, myth, and symbolism. In 1942, Gregory Bateson and Margaret Mead wrote of cockfighting in Bali, "The evidence of regarding the fighting cock as a genital symbol comes from postures of men holding cocks, the sex slang and sex jingles, and from Balinese carvings of men with fighting cocks." The British anthrozoologist Gary Marvin interpreted the Spanish cockfight as a celebration of manhood, and Princeton's Clifford Geertz argued that the function of rooster fights in Bali is to confirm the status hierarchy among the men in rural villages. More recently, in an essay titled "Gallus as Phallus," the University of California anthropologist Alan Dundes argued that a cockfight is a "homoerotic male battle with masturbatory nuances."

While I found these quasi-Freudian notions interesting, they provided no insight into the question I was interested in—how could such apparently normal people be involved in an activity that most Americans, myself included, viewed as an exercise in sadism? But to figure out the paradox of why seemingly good people do seemingly bad things, I had to learn about their sport. I had to step back and assume the mantle that the neurologist Oliver Sacks called an "anthropologist from Mars." Over the next two years, I racked up a couple of thousand miles driving the back roads of east Tennessee and western North Carolina, interviewing cockfighters, photographing their kids and roosters—usually together—and gathering data, checklist in hand, at clandestine rooster fights. Along the way, I learned a lot about how people think—and avoid thinking—about the way we treat animals.

THE COMPLICATED CULTURE OF COCKFIGHTING

Pitting roosters against each other is one of the oldest and most widespread of traditional sports. The chicken as we know it was domesticated from several species of wild Asian jungle fowl about 8,000 years ago. Wild jungle fowl are inveterate fighters, and humans were probably staging bouts between roosters from the time they began keeping chickens around for meat and eggs. The earliest recorded cockfights date back to 517 BCE in China and soon spread through Southeast Asia, the Pacific Islands, the Middle East, and eventually to ancient Greece and Rome, where young men were required to attend cockfights to learn the meaning of courage. The sport took hold in Europe and became popular in Spain, France, the British Isles, and Scandinavia. Columbus brought chickens–and probably cockfighting–to the New World, where it spread rapidly through North and South America.

To understand cockfighting, you start with the chickens. Cockfighters are obsessed with bloodlines. They talk endlessly about the merits of cross-breeding, linebreeding, and inbreeding. Some of them can tell you about F1 and F2 generations—just like your high school biology teacher. Breeders I visited would pull out notebooks that went back decades. They knew who sired whom and which hens produced good shufflers and cutters.

There are hundreds of strains of roosters, and they have great names—Blue Faced Hatches, Kelsos, Arkansas Travelers, Allen Round-heads, Madigan Grays, Butchers, Clarets. When breeding their battle cocks, rooster fighters are looking for the perfect combination of three traits. The first is cutting—the ability of a cock to deliver accurate strikes to the opponent's body, to puncture the lungs or heart. The second is the ability to put power behind the blows. But by far the most important trait, the one that gets breeders misty-eyed, is what they call true grit, or more commonly, gameness. I ask Johnny, a third-generation cocker, to tell me how I could explain gameness to my animal rights pals. "Gameness," he said, "is their heart. Their desire to fight to the death. Your barnyard rooster is cowardly. They can't take the steel of the gaff. Gameness is the drive to beat the opponent that is so instilled in the true game rooster that he is going to give everything he has to his last breath."

While cockers are as concerned with bloodlines as members of the Westminster Kennel Club, they will all tell you good genes are not enough to make a great fighting cock. You also have to raise them right. Johnny's stance on the poultry nature-nurture debate is that a rooster's fighting ability boils down to 85% genes and 15% conditioning. Two to four weeks before a derby, cockfighters put their roosters on a prefight diet called the "keep." Every cockfighter has their secret system. Johnny would give his roosters vitamin supplements and allow them to spend a few hours every day foraging for bugs on fresh grass. Some cockers start adding a pinch of strychnine to their rooster's food during the keep; they think it thickens a rooster's blood. Others dose their roosters with antibiotics, testosterone, or stimulants. Johnny tried Dexedrine a couple of times, but he gave it up when he realized the drug made his roosters act crazy in the pit.

Cockers think of their birds as athletes, and they have developed prefight conditioning regimes to develop stamina and quickness. Johnny worked his birds in the morning and then again in the afternoon. He had a padded exercise bench where he would put them on their backs so they learned to right themselves quickly, and he practiced "flirts" by flipping them backward to develop their wing and back muscles. Like iron pumpers at a Gold's Gym, Johnny's roosters got an exercise for each muscle group, and he kept track of each rooster's daily reps as he put it through its paces. In the weeks before a big derby, he would spend six hours a day conditioning his animals.

EVEN COCKFIGHTS HAVE RULES

Every sport has rules, and each of the world's cockfighting cultures has its traditions that govern how fights are conducted. In Andalusia, for example, metal gaffs are not attached to the roosters' spurs so Spanish cockfights are usually not lethal. The most common type of cockfight among southern cockers is called a "derby." In a derby, each cockfighter fights a preset number of roosters in a series of round-robin matches. In the United States, the basic set of guidelines is referred to as Wortham's Rules, in place since the 1920s.

The rules are complicated—you don't just put the roosters together and let them have at it. During fights, there are two handlers, two chickens, and a referee in the pit. The referee controls the action. The handlers, who often are not the actual owners of the birds, bring the roosters close enough to peck at each other's faces. Then they place the birds on the ground eight feet apart, facing each other. On the referee's command, the roosters are released. Instantly and silently, they charge toward each other in a flurry that Clifford Geertz described as "a wing-beating, head-thrusting, leg-kicking explosion of animal fury so pure, so absolute, and in its own way so beautiful, as to be almost abstract, a Platonic concept of hate."

Within twenty or thirty seconds, the gaffs usually become entangled in the body of one or both roosters, and they collapse to the ground in a heap. Then the referee signals the handlers to step forward and disentangle their birds. The handlers have twenty seconds to get their birds ready for the next pitting. A good handler knows the anatomy of injuries and, like the corner man in a prizefight, can bring a hurt animal back for the next round. He may put a little water on the cock's face, or blow on its head to settle him down, or just put him on the ground and leave him alone so the bird can sling a blood clot out of his throat. At the end of the rest period, the fight resumes. And the process repeats until there is a winner.

Sometimes a rooster will win a cockfight by killing its opponent outright or he will beat it up so badly that the opposing handler throws in the towel. But more often than not, the victor is determined by a complicated set of rules called the "count." Above all, cockers value a gamecock's drive to fight no matter if its lungs are punctured, its spine shattered, or its vision growing dim. The count system ensures that the gamer rooster, the one that keeps fighting when all seems lost, wins. When a rooster stops attacking because he is injured or exhausted, the opposing handler says to the referee, "Count me." If the other rooster does not attack for four successive pittings, the rooster whose handler "has the count" is declared the winner. But if the bloodied and battered rooster makes the slightest effort at aggression, even a weak peck toward its adversary, the count starts over. A typical gaff fight lasts about ten minutes, but sometimes a rooster will get in a lucky blow and hit a vital organ, and it is all over in seconds. On the other hand, a fight can go for an hour or more. To keep the spectators from getting bored, long fights are moved to a secondary or "drag pit" which frees up the main arena for a pair of fresh birds.

In recent years, the influx of immigrants has affected cockfighting just as it has many other aspects of American culture. The weapon of choice of the new breed of cockers is the knife. These artificial spurs, unlike icepick-shaped gaffs, have razor-sharp edges. Filipinos like "the long knife" which looks like something you would use to dismantle a porterhouse at a Ruth's Chris Steakhouse. Mexicans prefer the short knife which has a one-inch blade honed on both sides. Old-timers, like Johnny and Ed, aren't keen on these new developments. They claim the knife makes it easy for a less courageous rooster to win a fight by way of a lucky blow.

Cockfights are not nearly as bloody as you would imagine. When Eddy's wife, a college food service manager, describes the first time he took her to a fight at the Del Rio pit, she says, "I was surprised. It was not the blood and guts I thought it would be. And the restrooms were clean. So was the kitchen and the food was good. They made everything from scratch."

She is right about the gore. The wounds inflicted by the steel gaffs do not bleed much on the outside, and the cock's feathers conceal most of the leakage. And a game rooster's blood is more efficient at clotting than the blood of other types of chickens. Still, in the parlance of cockfighters, game roosters can "take a lot of steel." I once took the carcass of a rooster who had been killed in a fight to the state veterinary pathology lab for an autopsy. The pathologist, it turned out, had been involved in cockfighting when he was growing up in Oklahoma. When he opened the cock up, he counted nineteen holes in its body. The fatal blow was to the throat. Cockers have a lexicon of injury. A rooster that has had its lungs punctured emits a creepy rasping sound called the "rattles." When a cock flops around with spinal cord damage, it is said to be "uncoupled."

Fights are usually straightforward affairs; almost all the losers die and almost all the winners live. But every now and then, the unexpected happens. A rooster may refuse to fight and just walk around aimlessly with a bewildered look in his eyes. Occasionally a cock will humiliate its owner by turning tail and running around the pit, squawking. Then there are the man-fighters, cocks that spin around and go for their handler instead of the other rooster. One night I saw a Grey nail his handler in the thigh with both gaffs, two-and-a-half inches of ice pick. The man turned ashen and collapsed.

Notch one up for the chicken.

Gambling and cockfighting are inseparable. Money changes hands in two ways in derbies. Each of the cockers participating in a derby pays an entry fee, and the one whose roosters wins the most fights takes home the pot. In a typical small-time Appalachian pit back in the day, twenty or so cockers might enter a five-cock derby with a $200 entry fee, so the evening's winner would leave $4,000 richer. The second form of gambling at cockfights consists of informal side bets. The cockfighters in each match usually agree on a bet between themselves, but the real action is among the spectators. As soon as the handlers bring their roosters into the main pit, the people in the crowd start yelling out the odds they are willing to give on the birds. The odds are based on both the reputation of the breeder and the demeanor of the chickens. As soon as the winner is proclaimed by the referee, all bets are paid. Disputes over money are rare.

One of the biggest surprises of my forays into the clandestine world of rooster fighters was how open it was. In those days, Appalachian cockfighters made no real effort to hide their involvement in the sport even though it was against the law. As I drove around the region, I would see fields with hundreds of gamecocks, some tethered to barrels, others in little wooden houses that looked like Boy Scout pup tents—all of them in full view of the road, even interstate highways. How did they get away with it?

Simple. Back then, being a rooster fighter in Appalachia was about as illegal as being a litterbug. True, cockfighting was against the law in both North Carolina and Tennessee, but it was a misdemeanor, and local sheriffs generally turned a blind eye. On the rare occasions when raids occurred, the participants were invariably slapped on the wrists and given a $50 fine. No one ever went to jail. Sure, some sheriffs were on the take, but others figured it was better to have the fights in established pits where it was in everyone's interest to be on good behavior. These pits had implicit codes of conduct to avoid altercations—no drinking or drugs around the pit, don't welsh on bets, the referee's word goes. Pit owners also did their best to avoid trouble with their neighbors. The owner of the Ebbs Chapel pit would take donations at the Saturday night fights for the Baptist church up the road.

In short, the local law figured it was preferable to let the pits operate rather than force the sport more deeply underground. They may have been right. Cockfights involve a potentially explosive combination of testosterone and money. I was afraid only once at a cockfight, and it was at a ragtag event called a "brush fight." Brush fights are informal affairs held in barns or in the piney woods, arranged by a few last-minute phone calls, with no entry fee, no paid referees, and no rules against alcohol. The fight was in west Knoxville on a hot, late June afternoon and the beer was flowing. The referee was a spectator who was drafted from the crowd and did not know what he was doing. Two handlers, who had both been drinking, squabbled over one of the ref's calls. In a regular pit, they would have been ejected immediately. Not so in a brush fight. The argument escalated until one guy grabbed a beer bottle, broke the neck off, and went for the other handler, catching him in the shoulder. With blood dripping down his arm, the man who had been stabbed grabbed his chicken and stomped off.

When I heard him slur, "I'll teach that asshole a lesson. I've got a shotgun in the back of my truck," I started looking for an escape route. Unfortunately, I had gotten to the fight early and was hemmed in by a dozen pickup trucks. However, everyone else was as averse to drunks with shotguns as I was. Within a couple of minutes all the trucks had cleared out, and I was headed home, sweating, heart beating, and hands shaking, thinking that maybe I was not cut out to be an anthropologist after all.

WHO ARE THESE PEOPLE?

Most people imagine that cockfighters are low-life scum who peddle crystal meth when they are not gleefully torturing animals. My experience, however, was different. To me, the most psychologically interesting thing about rooster fighters was how boringly normal they were. Nearly all the cockers I met led—aside from their devotion to a brutal bloodsport—ordinary lives complete with mortgages, wives and children, and day jobs.

No one knows how many cockfighters there are in the United States. In the early 1990s, the sociologist Clifton Bryant estimated there were about 100,000 active participants, but that number is fuzzy and has plummeted in recent years. Increasingly stiff penalties at the state and federal level have caused once avid cockers to trade in their gaffs for bass boats and golf clubs. Fighting roosters or attending a cockfight is now a federal crime. And while cockfighting has been called Puerto Rico's "national sport," it was banned on the island when President Trump signed amendments to the 2018 Farm Bill.

The people I encountered in my research were representative of Appalachian rooster fighters. Willy owned a Shell station, Frank was the mayor of a small town, and Larry a long-haul trucker. Richard was a police detective who only quit cockfighting when the higher-ups in the department dangled a big promotion in front of him on the condition that he give up his chicken habit. Most of them were hobbyists who kept two or three dozen battle cocks in their backyards and fought in a handful of derbies every year. On the other hand, Clyde Robinson, who lived near Rosman, North Carolina, and Tim Davenport of Sparta, Tennessee, were internationally known breeders who shipped hundreds of roosters every year to places like Hawaii and the Philippines. For them, cocking was a family affair. Their wives and kids helped with the husbandry. Mr. Davenport's fourteen-year-old son was developing into a deft handler who regularly won against men old enough to be his grandfather.

Suzie, a Louisiana animal protectionist, worked tirelessly for years to ban cockfighting in her state. Yet her experience with rural southern rooster fighters was similar to mine. She despises cockfighting. But, like Baptists who claim to hate the sin but not the sinner, she came to respect some of her opponents. She told me, "Most of the cockfighters I met were God-fearing, polite family people who were not out there shooting heroin or snorting coke. I am completely opposed to cockfighting, but that does not mean that cockfighters are evil people."

JUSTIFYING THE UNJUSTIFIABLE

If cockfighters were sadistic perverts, it would be easy to explain their involvement in a cruel bloodsport. But most of them are not. So, how can they participate in an activity that is illegal and that nearly everyone in America thinks is immoral? The answer is that they construct a moral framework based on a mix of wishful thinking and twisted logic in which cockfighting becomes completely acceptable. In this regard, they are no different from any other person who exploits animals—hunters, circus animal trainers, even scientists and meat-eaters. There are several lines of argument they use to justify an activity that most people find unjustifiable.

It's the Most Humane Sport

In his book *Fighting Sports*, Captain L. Fitz-Barnard claimed, "Cockfighting is the most humane, perhaps the only humane, sport there is." While his claim is extreme even among avid cockers, most cockfighters deny that their sport is cruel. They tell you that the fight itself is only a small part of the sport. They tell you it takes two years to raise a chicken from hatchling to battle cock, and that the fight, which is often over in a few minutes, is just a fraction of what it means to be a cocker.

But what about the pain? My friend Johnny claimed that the steel gaffs had taken the cruelty out of cockfighting. The gaffs, he argued, make it a fair fight because they equalize each rooster's chance of winning. If it were not for the gaffs, he says, the roosters would batter each other to death with their three-inch natural spurs.

My neighbor Paul Ledford laid another argument on me one morning when I asked him in his kitchen over a cup of coffee how much pain was inflicted during a fight. He shook his head and said, "Chickens don't feel pain. Chickens are too dumb to feel pain." And Richard, the police detective, told me that a chicken's brain is "wired differently from the human nervous system."

You also hear the opposite argument, that chickens are free moral agents who choose to fight each other to the death. According to this thread of logic, it is cruel NOT to let roosters meet their destiny in the cockpit. Fitz-Barnard wrote, "Where the agents are willing, there can be no cruelty. For the gamecock, the joy of battle is his greatest joy." Fitz-Barnard thought that cockfighting is ethically superior to hunting or fishing because the whitetail buck or brown trout you kill does not have a choice in the matter. He wrote, "Where you have unwilling participants there must be cruelty . . . no sane person can pretend that the fish enjoys being lured to death, often with live bait; that the fox or the hare likes to be hunted and torn to pieces; or that the birds and beasts prefer a lingering death from gunshot wounds."

I don't buy for a minute the argument that gamecocks choose to fight because they achieve the avian version of self-actualization in battle. No, they fight because their brains are hard-wired by thousands of years of intense selection to plunge their spurs into the bodies of other male roosters. Even if they wanted to flee, the close quarters of the pit make it impossible. Fitz-Barnard's comparison of cockfighting and hunting, however, does strike an uncomfortable chord. As many as 30% of the 120 million wild birds shot by hunters each year in the United States will fall from the sky wounded, yet conscious. The lucky ones will be found and killed quickly, but millions more will die lingering deaths. Fitz-Barnard is right: There is much more suffering caused by the legal sport of recreational hunting than the outlawed sport of cockfighting.

It's Only Natural

A common variation on the "it's not cruel" justification for cockfighting is that roosters are instinctive fighters, just like lions are instinctive zebra killers. This is a variation on the well-known naturalistic fallacy. Here is how Johnny laid it out for me: "What we do is an act of nature in a controlled situation. That rooster's going to fight if we are there or not. We make things as even as we can for them to perform an act of nature. We make sure there isn't a difference in their size or weights or the shape of their spurs. We don't make these roosters fight. That's what they were put here for. It is their purpose." (This, by the way, is also the reason most cockfighters I met did not approve of dogfighting. As Eddy's wife told me, "Cockfighting is not like dogfighting. They have to make the dogs mean. But these chickens are born to fight. They do it regardless of whether you are there or not.")

I run across the naturalistic fallacy a lot when I talk to people about the use of animals. In explaining her opposition to animal research, an animal rights activist friend told me that AIDS and swine flu were "nature's way" of reducing human overpopulation. A woman once told me at a party, "I just don't get vegetarians. Humans have been eating animals for millions of years. That's what cows and chickens were put here for." I did not point out that her justification for eating chickens was identical to Johnny's rationale for fighting them. I suspect she would not have seen the parallels.

The Nicest People You Ever Saw

Restrictions on cockfighting were originally based not on concern for animal suffering, but for keeping the riffraff down. The association between cockfighting and other forms of rowdiness continues today. For example, the Humane Society of the United States has linked cockfighting with murder, prostitution, identity theft, robbery, Mexican drug cartels, the manufacture and sale of narcotics, illegal gambling, bribery, gang activity, tax evasion, money laundering, immigration violations, and the possession of weapons including knives, guns, and hand grenades.

Cockfighters, of course, don't see it that way. They view themselves as a misunderstood band of brothers held together by a common set of values that include hard work, competition, respect for cultural traditions, and love of chickens. They dismiss the charges of booze, dope, hookers, and money scams. Of animal protection organizations, Johnny told me, "They call us everything—pimps, drug dealers. To them, we are the scum of the earth. They are smarter than we are. They know how to paint us to people who don't understand." OK, he admitted, there are a few bad apples—the cheat who dabs poison on his rooster's hackles or illegally sharpens the edges of a long-heel gaff. But they are a small minority. As for 99% of cockers, he said, "You will never find a nicer group of people. Where else can you go where that much money changes hands on a nod without any controversy? Cockers are gentlemen."

The Great Man Defense

Cockers supplement the "good people" defense with a rhetorical twist social psychologists call *reflected glory*. The logic goes like this: "If _____ was a cockfighter, it must be OK." The list of names they insert in the blank includes George Washington, Alexander Hamilton, John Adams, Alexander the Great, Woodrow Wilson, Andrew Jackson, Henry Clay, Hannibal, Caesar, Thomas Jefferson, Benjamin Franklin, and Abraham Lincoln, who was reportedly a referee. Genghis Khan and Helen Keller are sometimes added to the list, though I am skeptical about Helen Keller. A long line of British royalty are claimed to be fellow travelers by modern cockfighters who often refer to rooster fighting as the "sport of kings."

Cockfighting Builds Character

When Bobby Keener of Greensboro, North Carolina, was asked what kept him interested in cockfighting, he replied, "It's the way that this animal would keep going until he has nothing else to give. How many people would do that? This bird will give you everything he's got until he's got no more and then he keeps giving it. It's what you call gameness or heart. That's what's kept me interested in it." I call this the moral model defense. To a cockfighter, a game rooster is the bravest creature on earth. That's why the gamecock is the mascot of the University of South Carolina football team. A cocker summed up the moral model argument when he wrote in the cockfighting magazine *Grit and Steel*, "A game-cock is loyal to his family and himself—and he has the grit to back that loyalty. . . . It takes grit to be loyal—to your ideals, to your wife, to your husband, to your friends, to your country."

I Love My Chickens

For a chicken, a game roster has a pretty good life. Cocks are not usually fought until they are two years old, during which time they lead a life befitting a thoroughbred racehorse. For the first eight or nine months, they have the run of the chicken yard. Once they hit puberty, the cocks have to be separated, but because breeders want their birds to get exercise, they are either tethered with seven-foot "tie out" cords or kept in large cages that allow them to move about. In addition to the organic corn that Johnny bought for them at the health food store, his roosters got hard-boiled eggs for breakfast and dined on fruit, greens, and pearl barley for lunch. Every other day, he would give them a little hamburger meat supplemented with cottage cheese. Johnny complained to me, "We give our roosters the best food, the best housing, the best hens. . . . And then they call us cruel."

Like every cockfighter I have ever talked to, Eddy Buckner was passionate about his roosters. He told me he loves them. And I believe him. Like Fabe Webb, his eyes sparkled when he talked about his birds.

"But Eddy," I said. "You claim to love these animals. You raise them practically by hand for two years, you spend hours every day handling them and exercising them. But then you take them to the pit on the weekend knowing full well that half of them are going to die before the end of the night, and then you just toss them in a barrel. I don't get it."

"You have to draw the line," he said.

"But don't you get attached to them?" I asked.

"Sure," he said.

"Do you ever name them?"

"Yes."

"Did you ever see a cockfighter cry over a dead rooster?"

"Never."

"I still don't get it," I said.

ANIMAL ACTIVISTS VERSUS ROOSTER FIGHTERS: AN ASYMMETRY OF HATE

Cockfighters are completely convinced by their ethical arguments—so convinced that rooster fighters have told me with a straight face they would like to bring animal rights activists to a derby so they could see what a wonderful sport cockfighting is. This idea is ludicrous. Every animal protectionist thinks that cockfighting is cruel, and nothing would ever change their minds. This leads to an asymmetry of animosity—animal activists hate cockfighters more than cockfighters hate animal protectionists.

Karen Davis has spent her life working tirelessly for the protection of animals. She is the founder of United Poultry Concerns, the country's only poultry rights organization, and she despises rooster fighting. She says that the sport is not about competition, but about male insecurity. The great irony of manhood, she told me, is that men are scared of each other, that they are afraid that other men will see their weakness, their touch of female sensibility. To her, cockfighting is a perverse adult version of the playground taunt, "My dad can beat up your dad."

When I asked her how she feels about cockfighters, Karen began to work herself up. "A chicken is about as tall as your knee, and they are at the total mercy of these cockfighters who tower over these little birds and punish their bodies. And the ultimate punishment is putting them in a ring and making them beat each other up. These people force their violent bloodlusting impulses on birds. And then they say they admire these animals and love them. I don't think there is a chicken in the world that would be grateful for that."

Is she right? If given a choice, would any chicken in the world volunteer to be a gamecock? Karen Davis forced me to revisit the question I asked myself while I was chasing down the chicken trucks on I-40: Would I rather be a fighting rooster or a commercial broiler? To answer it, we need to look at the sorry lot of the industrial chicken.

WHAT WOULD YOU RATHER BE, A GAMECOCK OR A BROILER?

The modern broiler is a technological marvel. The chickens I followed down the interstate looked to be Cobb 500s, one of the world's most popular types of broiler. The Cobb 500 was developed by the Cobb-Vantress company, a multinational corporation formed in 1986 as a joint venture between two corporate giants, Tyson Foods and Upjohn, the pharmaceutical company. Cobb-Vantress, which has operations in Europe, Asia, South America, and Africa, also produces the Cobb 700, designed for a high proportion of breast meat; the Cobb Sasso 150, aimed at the free-range and organic market; and the Cobb Avian 48, which is advertised as having "high livability" and particularly suitable for "live bird markets seen in some parts of the world."

These birds are meat machines. The average Cobb 500 breeder hen will produce 132 chicks by the time she is "depleted" at the age of fifteen months. Her chicks will have a much shorter life than their mother. In 1925, it took 120 days and ten pounds of feed to produce a scrawny two-and-a-half-pound bird. Now the optimal slaughter age of a modern broiler is thirty-nine days when they will weigh between four and five pounds. While a modern chicken grows nearly five times as fast as your grandmother's barnyard chicken, it eats way less food. In 1925, it took nearly five pounds of feed to produce a pound of chicken. Now, the conversion ratio of feed to meat conversion ratio is 1.6 to one. Even better from an industry perspective, there is little waste on a modern chicken. Once its feathers are plucked, its feet and head chopped off, its guts scraped out and blood drained, 73% of a Cobb 500's carcass will be "eviscerated yield."

Cheap meat comes at a cost. A broiler chicken's bones cannot keep up with the explosive growth of its body. According to the Department of Agriculture, if a human grew as fast as a chicken, an eight-week-old infant would weigh 350 pounds. Unnaturally large breasts torque a chicken's legs, causing lameness, ruptured tendons, and "twisted leg syndrome." Donald Broome, a professor of animal welfare at Cambridge University, told me he believes severe leg pain in chickens is the world's largest animal welfare problem. Arthritis, heart disease, sudden death syndrome, and a host of metabolic disorders are prevalent among industrial broilers.

The living conditions of the animals destined to become chicken tenders are Dante-esque. The chicks will never see sun or sky. Because they are so top-heavy, broiler chickens spend nearly all day lying down, often in litter contaminated with excrement. As a result, many will develop breast blisters, hock burns, and sores on their feet. A chicken "grow-out house" can be 600 feet long and 60 feet wide and hold 30,000 birds. The broiler houses are dark and humid, and the air is laced with ammonia produced by the action of microbes on accumulated urine and excrement of tens of thousands of birds. The gas burns the lungs, inflames the eyes, and causes chronic respiratory disease.

When the broilers reach four or five pounds, it is time for their trip to the processing plant. But first, you have to catch them. Crews of low-paid chicken catchers invade the sheds at night when the birds can't see. Wearing masks and throwaway coveralls, they grab the chickens by their feet, five birds in each hand, and stuff them into holding drawers. It can take all night for a team of catchers to load 30,000 chickens into wire crates, during which time each catcher will lift fifteen tons of flapping, squawking birds. It is nasty work for people and animals. The catchers get scratched, pecked, and covered with shit. As many as 25% of the birds will be injured in the catching and loading process. My colleague Bruce Henderson, a developmental psychologist, signed onto a chicken catching crew when he was in high school. He lasted six days.

Isn't there is a better way to snatch up chickens? Enter the mechanical chicken catcher. There are several varieties of these behemoths. One of the most popular ones is the Apollo Generation 2 mechanical chicken harvester produced by CMC Industries. It works more or less like a pair of giant dustpans. In the dark, the machine's "harvesting heads" gingerly sweep the two metal ramps through the flock of chickens. They nudge each other onto the ramp where the birds are moved toward a conveyor belt that transports them directly into wire transport cages for the trip to the processing plant. According to the company, "Birds are loaded in a natural, stress-free manner, gently deposited in the drawers, with no direct contact from operators." The machine is a labor saver. It can sweep up 12,000 birds in an hour with only three operators.

The Humane Society of the United States has claimed that mechanical harvesters are easier on birds than the conventional method of hand catching. Chickens hate being grabbed by humans and held upside down by their sore legs. They seem less stressed as they march clueless into the maw of a giant machine and excreted moments later, a bit dazed, into wire transport cages. The Lewis/Mola Company claimed their mechanical harvester produced 60% fewer broken wings and a 99% reduction in broken legs. Despite these presumed advantages, the vast majority of broilers consumed each year in the United States are still captured by hand.

Once the transport cages are loaded, the trucks hit the road and our Cobb 500s are on their way to one of the 160 industrial poultry processing plants in the United States where they will be dumped from their crates, shackled tightly by their legs with metal cuffs, and hung head-down from a conveyer belt. Then, twenty at a time, upside down and wings flapping, their heads will be passed through an electrified water bath. For seven to ten seconds, electricity will travel through their bodies, hopefully stunning the animals.

The next stop is the neck-cutting machine where a set of rotating blades will sever the chickens' carotid arteries. Once the chickens bleed out, their bodies will be dunked in the scald tank. In 2021, in response to industry pressure, the United States Department of Agriculture increased the maximum evisceration speed in poultry houses from 140 birds per minute to 175 birds. Even when birds are killed by hand, as in the case of kosher slaughter, one worker should be able to slit the throats of 4,000 birds per hour. While efficient, this system does not always work. Some birds are not completely stunned before their throats are slit, and if the neck-cutting blades miss a chicken's carotid artery, the animal will be conscious when it is submerged in the hot water of the scald tank. Then, after the birds are killed and eviscerated by the machines, the assembly line workers take over. Their official job titles speak for themselves: vent opener, neck breaker, giblet harvester, gizzard table peeler, heart and liver cutter, lung vacuumer, and, finally, bagger.

There you have it. Your average east Tennessee gamecock chick will be pampered during its two-year life. For the first six months, it will run free. After that, it will have 150 square feet of lawn to loll about, and it will have a private bedroom to sleep in at night. The rooster will get plenty of exercise and eat better than some people. It will have a chance to chase the hens around. The downside is that one Saturday night, he will feel the pain of the Mexican short knife slicing his pectoral muscles or perhaps get a long heel gaff in the throat, and he will die in the dirt in a fight that could last anywhere from a few seconds to over an hour while men sporting baseball caps shout out odds to each other. His chances of seeing the sunrise on Sunday morning are fifty-fifty.

In contrast, our Cobb 500 chick will live in unimaginable squalor, legs aching, lungs burning, never glimpsing the sky or walking on grass or having sex or pecking at a bug, eating the same dreary industrial poultry chow day after day for its short forty-two-day life, when it will be jammed into a crate on an open truck and carted to the plant where it will be suspended upside down, electrocuted, and its throat slit. His chances of seeing the sunrise are zero.

Karen Davis tells me that no chicken in the world would want to live the life of a fighting rooster. I'll lay 25-to-20 that she is wrong.

HOW MONEY AND SOCIAL CLASS AFFECT OUR PERCEPTIONS OF CRUELTY

Objectively, it is hard to deny that much less animal suffering is caused by cockfighting than by our apparently insatiable demand for chicken flesh. Even back in the 1980s, 10,000 or 20,000 chickens had their necks slashed in a mechanized processing plant for each gamecock who died in a derby. Then there is the inconvenient fact that the life of a fighting cock is fifteen times longer and infinitely more pleasurable than the life of a broiler. Why, then, is it legal for us to kill more than nine billion broiler chickens every year but cockfighting can get you five years hard time in a federal penitentiary plus a $250,000 fine?

It is, in part, a matter of money and power. Founded in 1954, the National Chicken Council is the trade association of the $70 billion poultry industry. Its members include Tyson Foods, Pilgrim's Pride, and Purdue Farms—the corporate giants that produce 95% of the broilers consumed in the United States, and the organization works tirelessly to keep the government at arm's length. As a result, factory-farmed chickens are exempt from virtually all federal animal welfare statutes including the Humane Methods of Slaughter Act, enacted by Congress in 1958 to ensure that the animals destined for the table will not unduly suffer as they are being killed.

While the National Chicken Council furthers the interests of the broiler industry, organizations such as United Poultry Concerns, PETA, the Farm Sanctuary, and the Humane Society of the United States (HSUS) are the voices of America's chickens. Of these, the HSUS has the most political clout. With assets valued at over $330 million, the HSUS is the 800-pound gorilla of the animal protection movement.

In 1998, the HSUS decided to take on the cockfighters. In response to political pressure from animal protection groups, the last states caved. Louisiana was the final holdout. As the state sought to improve its image in the wake of Hurricane Katrina, the influence of the cockfighting lobby in the legislature waned, and in 2007, Louisiana Governor Kathleen Blanco signed a bill that closed the loophole that excluded chickens from the state's animal cruelty laws. State cockfighting laws remain inconsistent. In Florida, the penalty for cockfighting is five years in jail and a $5,000 fine, but in neighboring Alabama, it is a $50 misdemeanor. Now that cockfighting is banned everywhere, HSUS's mission is to rally their ten million members to pressure lawmakers to make cockfighting a felony in every state. And the HSUS offers a $1,000 reward to anyone who reports someone for being involved with cockfighting.

The war on cockfighting is about cruelty, but the subtext is social class. The eighteenth-century movement against bloodsports was directed toward activities like bull-baiting and cockfighting that appealed to the proletariat rather than leisure pursuits of the gentry such as fox-hunting. It's no different today. Cockfighters come from easy groups to pick on—Hispanics and rural working-class whites. Animal activists, on the other hand, tend to be urban, middle class, and well educated. They dismiss rooster fighters as a motley amalgam of ignorant shit-kickers and illegal aliens.

Duke University's Kathy Rudy, an avid animal advocate and dog rescuer, is troubled by the divide between animal activists and the working class. In an essay that appeared in the *Atlanta Journal-Constitution* shortly after former Atlanta Falcons' quarterback Michael Vick pled guilty to charges of dogfighting, Rudy pointed out that our society is much more likely to criminalize forms of animal abuse that involve minorities and the poor than animal cruelties that affect the wealthy. The comedian Chris Rock made the same point on national television in response to a photograph of Sarah Palin, the governor of Alaska, that had gone viral. An avid big-game hunter, she was running for vice president of the United States. He told David Letterman, "She's holding a dead, bloody moose. And Michael Vick's like, 'Why am I in jail?' They let a white lady shoot a moose, but a black man wants to kill a dog? Now that's a crime."

You see the same thing in thoroughbred racing. About 1,000 horses a year die on American racetracks. And that number would probably be doubled if you included the horses who die in private training facilities. Yet, according to a recent national poll, only 20% of Americans would support a ban on thoroughbred racing. Like cockfighting, horse racing represents a confluence of gambling and suffering. But unlike cockfighting, thoroughbreds are the hobby of the rich and famous.

COCKFIGHTING AND HUMAN MORALITY

I am conflicted about many moral issues involving animals. Cockfighting, however, is not one of them. I liked most of the cockfighters I met in the course of my research, but, as was the case with slavery, their sport is a cruel and unjustifiable anachronism. It is time for rooster fighters to close down the pits and spend their Saturday nights home with the family.

Yet I am still disturbed about attitudes about cockfighting and what they say about moral hypocrisy and the fallibility of common sense in our relationships with other species. While the great chicken-eating public (of which I am a member) will sleep easy tonight knowing cockfighting is now banned in every state, teams of chicken catchers from Maryland to California will enter darkened broiler houses and stuff thirty-five million terrified birds into wire crates in preparation for their journey to the processing plant tomorrow.

Back when I was doing my doctoral research on cockfighting, I took an organizer from Amnesty International named Tony Dunbar to a derby at the Ebbs Chapel pit. Tony's day job was saving convicted murderers from execution. As we were driving home at 2 A.M. reeking of tobacco smoke and greasy cheeseburgers, I asked him what he thought of our evening rubbing shoulders with some of western North Carolina's ace rooster fighters. After a pause, he said, "To me, cockfighting is a small moral problem."

A lot of people would disagree. But I find it hard to escape the conclusion that, in comparison to the suffering that goes into producing a six-piece Chicken McNugget Happy Meal, he was right.

Delicious, Dangerous, Disgusting, and Dead

THE HUMAN-MEAT RELATIONSHIP

> The pig has a stronger interest than anyone in the demand for bacon.
>
> —PETER SINGER, MORAL PHILOSPHER

> All normal people love meat. You don't win friends with salad.
> —HOMER SIMPSON, CARTOON CHARACTER

Staci Giani is forty-one but looks ten years younger. Raised in the Connecticut suburbs, she lives with her partner, Gregory, in a self-sustaining eco-community deep in the mountains twenty minutes north of Old Fort, North Carolina. Staci radiates strength, and when she talks about food, she gets excited and seems to glow. She is an Italian-American, attractive, and you want to smile when you talk to her. She tells me that she and Gregory built their own house, even cutting the timber and milling the logs. I think to myself, "This woman could kick my ass."

Staci wasn't always so fit. In her early thirties, Staci's health started going downhill. After twelve years of strict vegetarianism, she began to suffer from anemia and chronic fatigue syndrome, and she experienced stomach pains for two hours after every meal. "I was completely debilitated," she tells me. "Then I changed the way I ate."

"Tell me about your diet now. What did you have for breakfast today?"
I ask.

"A half-pint of raw beef liver," she says.

■ ■ ■

Animal activists sometimes claim that Americans *en masse* are forsaking barbecued ribs and buffalo wings for garbanzo bean burgers and tofu nut loaf. True, an increasing number of people believe that animals are entitled to basic rights, including, one presumes, the right not to be killed because you happen to be made out of meat. But despite our stated love for animals, Americans eat seventy-two billion pounds of animal flesh each year and only a small fraction of people in the United States are true vegetarians. We kill 500 food animals for every animal used in a scientific experiment, 15,000 for each unwanted dog euthanized in an animal shelter, and 150,000 for every baby harp seal bludgeoned to death on a Canadian ice floe. And, in spite of what you sometimes hear, over the past thirty years, the animal rights movement has not made much of a dent in our desire to eat members of other species.

In most cultures, meat is a symbol of wealth, and as a nation gets richer, its citizens want to eat more of it. Since the 1980s, for example, meat consumption in China jumped from thirty-eight pounds per person to 140 pounds. Frank Bruni, former *New York Times* restaurant critic, captured the luxury of meat when he described a $90 steak he had for dinner one evening at a Manhattan chophouse. It was, he wrote, "a sublime hunk of glorious meat that you dream about for hours later, pine for the next day and extol in a manner so rapturous and nonstop that friends begin to worry less about your cholesterol than your sanity."

My revelation about the transcendent pleasures of meat occurred when Mary Jean and I were treated to an obscenely expensive dinner by a couple of old friends who were celebrating a milestone in their lives. Two waiters for our table; five wines, selected by the sommelier to match the food; a teaspoon of lemon shaved ice to perk up the taste buds between the soup and the fish course. The entrees were up to the chef, but there were a couple of appetizer options. Mary Jean chose duck confit. I went for the pork belly.

I had never tasted pork belly, but I remembered back when a local country music station would announce the going price during the noon farm and home show. The pork belly on the plate in front of me was a no-frills chunk of braised fat. One bite and my ideas about meat changed. I once stood for ten minutes in a museum staring at a Mark Rothko painting trying to figure out why anyone would think that an all-black canvas was art, but then something clicked and I suddenly got it. I had the same response to the taste of that pork belly. The Rothko and the pork belly had the same Platonic purity. One was the essence of blackness, the other of distilled meat.

What is it about meat that brings out the contradictions and conflicts in our thinking about other species? The problem is that while meat tastes good, it can be unhealthy, it is disgusting, and it involves killing animals.

WHY IS MEAT SO TASTY?

When asked in a Gallup poll what they would choose for their "perfect meal," twenty times more people said they would pick a meat entrée than said they would go for a main dish made of plants. Even half of vegetarians admit that they sometimes have meat cravings. Why are humans so drawn to the taste of flesh? The reason is simple—we evolved from a long line of meat-eaters.

Chimpanzees, our closest living relatives, love the taste of meat. Richard Wrangham, the Harvard University primatologist who has studied chimps for decades, says he has never heard of a wild chimpanzee who was not crazy about meat. The evolutionary anthropologist

Craig Stanford has seen juvenile chimpanzees pathetically staring up at meat-eating adults sitting in a tree, desperately hoping that a few drops of blood will fall their way. While adult chimps hunt rats, squirrels, small antelope, baboons, and even baby chimpanzees, their favorite meal is typically red colobus monkey. Chimpanzee carnivory is brutal and, sometimes, cannibalistic. The chimps of Taï National Park in the Côte d'Ivoire kill by disembowelment, while the Gombe chimpanzees of Tanzania are prone to torture their victims, tearing their limbs off or smashing their heads against tree trunks or rocks. In the Kibale forest of Uganda, chimps usually begin their meal by eating the viscera and organs of their prey, while it is still alive. And in some chimp communities, mothers occasionally eat their own infants.

Oddly, while chimps enjoy flesh, they don't eat all that much of it. Meat makes up only 3% or 4% of a typical chimpanzee's diet, and even the most voracious chimpanzee meat-eaters consume a couple of ounces a day. By two and a half million years ago, our hominid ancestors were eating more meat than modern chimpanzees. This shift to being a true omnivore was accompanied by changes in body and brain. Compared to apes, modern humans have a relatively small gut, with less devoted to the colon and more to the small intestine. Our teeth also reflect a fleshier diet. Like chimpanzees and gorillas, the teeth of our early *australopithecine* forebears lacked the concave shearing blades of serious meat-eaters; their large flat molars were designed to masticate tough vegetation and grind up husks of seeds and nuts. Our teeth, in comparison, are a Swiss Army knife of slicers, crushers, and biters.

The most important effect of omnivory was on the evolution of the human brain. Many anthropologists believe the shift to eating meat was a critical factor in the tripling of the size of our brains over a couple of million years. The idea that we evolved from meat-hungry apes is not new. Raymond Dart, who in 1924 discovered the first "ape-man" fossil, the Taung Child, in South Africa, wrote that our ancestors were blood-thirsty killers who delighted in "greedily devouring livid writhing flesh." Recent theories about the role of meat in human evolution are more sophisticated than Dart's, but the basic idea is the same. Craig Stanford argues that it is not a coincidence that humans, chimpanzees, baboons, and capuchin monkeys—the primates who eat the most meat—are also the most Machiavellian social manipulators. They are good at deception and making alliances; they are skilled in the nuances of complicated interpersonal relationships. He believes that meat jump-started the evolution of the human brain by facilitating social intelligence.

Our forebears probably made several dietary shifts, first from a high-fiber diet based on plants to a diet that included more animal flesh and then, after the invention of agriculture, back to more vegetable-based fare. The diets of hunter-gatherer peoples show that humans can adapt to an extraordinary range of foods, but they always included some meat. Before snowmobiles and satellite TVs came to the Arctic Circle, 99% of the caloric intake of the Nunamuit people of northern Alaska was derived from animal products—foods like raw *muktuk* (whale skin and blubber), fish, walrus, and fermented seal flipper. On the other hand, the !Kung of the Kalahari Desert survived easily on a diet composed of 85% plants. After surveying the diet of hundreds of hunter-gatherer groups, Loren Cordain of Colorado State University found that these groups obtained, on average, two-thirds of their calories from animal flesh, and not a single hunter-gatherer group survived on a diet composed of less than 15% animal products.

BoBo Lee, whose real name is Robert E. Lee (seriously), agrees with the theory that meat-eating is part of the natural order of things. For many years, BoBo ran Po-Pigs BBQ, a small barbecue joint next to a gas station forty-five minutes outside of Charleston, South Carolina. BoBo was a commodities trader before he became a barbecue man. I stumbled upon Po-Pigs while driving around what South Carolinians call the Low Country. I discovered later that it had been named one of the top five barbecue places in the United States.

In my experience, most barbecue restaurants disappoint. Not Po-Pigs. The atmosphere was right—checkered plastic tablecloths, a hand-printed sign taped on the door saying "cash and checks only," and, most importantly, no windows (always a good sign in a barbecue place). The pulled-pork was sublime—moist, smoky, and sweet. While there were four squeeze bottles of sauces on each table, they were distractions; the meat spoke for itself. When I complimented BoBo, he told me to come back the next morning and we could talk meat. I showed up at 9 A.M. and George Green, a retired army mess sergeant with a heavy Gullah accent, took me into the backroom and opened the top of the cooker. Through the smoke, I made out a couple of dozen mahogany-colored bone-in pork butts. They had been cooking all night at 200 degrees and were ready to be taken off the heat. After letting them sit for a half-hour, BoBo and George donned insulated gloves and pulled the hot meat apart by hand, mixing in a little homemade Pee Dee vinegar-pepper sauce. BoBo told me that if you have to chop up the pork with a knife, you haven't cooked it slowly enough.

I asked BoBo why humans find meat so satisfying.

"That's how our ancestors survived," he said. "By killing animals. Hell, they even ate mastodon. It's been ingrained in our heads that sitting down and eating a good piece of meat is a sign of success. It makes your mind feel good; it makes your stomach feel good. There is nothing as satisfying as a great warm, bloody, medium-rare steak."

But that does not explain why his wife, Pam, would not touch the stuff. She told me she didn't even like to look at raw meat. I asked her, "You have served a couple of hundred pounds of pork a day for the last ten years, but you don't eat meat?"

"I never liked the taste of meat," she said. "Even when I was a child my mother would put it in my mouth, but I wouldn't swallow it. It was not so much the flavor; I didn't like the texture. Now they make a vegetarian hotdog so I am in hog heaven."

"How do you like the barbecue here?"

"I have never tasted it."

NECESSARY, NICE, NATURAL, NORMAL . . . AND NASTY

Quoting F. Scott Fitzgerald, the cartoon character Homer Simpson once muttered to his wife, Marge, "The test of a first-rate intelligence is the ability to hold two opposing ideas in the mind at the same time, and still retain the ability to function." Homer's observation certainly applies to the fact that most people simultaneously love animals and love to eat them. Psychologists call this the "meat paradox." The cockfighters I studied developed convoluted sets of rationalizations for their involvement in a brutal bloodsport. The same is true of meat-eaters. In her book *Why We Love Dogs, Eat Pigs, and Wear Cows*, the social psychologist and animal activist Melanie Joy wrote that using animals for food rests on three arguments—meat-eating is normal, natural, and necessary. Jared Piazza and his colleagues added nice. Collectively, these are referred to as the "4Ns."

I would argue that another "N" also colors the human-meat relationship—nasty. Meat is the most dangerous of the foods we eat. When ABC News surveyed Americans about the kinds of food they were most afraid of getting sick from, 85% of people listed a form of meat compared to 1% who mentioned a vegetable product.

Our fear of animal flesh is well-founded. The problem is that we are made of meat, and thus susceptible to the various bacteria, viruses, protozoans, amoebas, and parasitic worms that can infect the creatures we eat. Fish carry at least fifty kinds of transmittable parasites. The flesh of cows, pigs, goats, and sheep can potentially carry the bacteria *Escherichia coli* which causes 400,000 human deaths worldwide each year. Some researchers believe that the AIDS virus was first transmitted to humans through the practice of eating chimpanzees, and others argue that COVID-19 originated in Chinese wildlife markets. Then there are the insidious little prions—tiny pieces of protein that have the remarkable ability to reproduce in your cells even though they contain no genetic material. They are responsible for mad cow disease which slowly turns the brain into Swiss cheese. Prions also cause kuru, a neurological disease found among the Fore people of New Guinea which is transmitted by the consumption of the brain tissue of deceased relatives during funeral rites.

But, you ask, why don't lions and wolves get sick from eating raw meat? Harvard's Richard Wrangham chalked it up to cooking. He argued that the invention of cooking by *Homo erectus* two million years ago was the breakthrough that made the big brain possible by opening a wider range of edible foods. In addition to adding flavor to their impala tenderloins, cooking also destroyed many of the pathogens that could make our ancestors ill. As a result, humans did not need to evolve the biological defense mechanisms that enable true carnivores to resist the toxins produced by the bacteria that live on meat.

The addition of spices to our diet also made meat less dangerous. The evolutionary biologist Paul Sherman wondered why humans are unique among animals in spicing the food we eat, particularly with inherently aversive substances like red hot chilies, which can burn you at both ends. Sherman hypothesized that humans developed the taste for spicy foods because spices contain chemicals that retard the growth of harmful microbes. To test this idea, he analyzed thousands of traditional recipes from all over the world. In every country, meat dishes were spicier than vegetable dishes. Further, people living in hot and humid areas more conducive to the growth of bacteria eat spicier foods than people in cooler climates. In countries like India, Indonesia, Malaysia, Nigeria, and Thailand, every meat dish is heavily spiced. (By the way, the most potent antibacterial spices are garlic, onion, allspice, and oregano.)

While cooking and spicing may make meat safer, eating flesh is still risky, especially for pregnant women. Meat-borne pathogens such as *Toxoplasma gondii, Listeria monocyogenes, E. coli, Shilella dysenteriae,* and *Leptospira* can cause spontaneous abortions, stillbirths, and prematurity. But evolution has come up with defenses against foods that can harm a developing embryo—nausea and food aversions. Most women experience nausea and vomiting during the first three months of pregnancy. That is when their developing embryos are most susceptible to the effects of toxins. Sherman reasoned that because meats are the most dangerous of foods to eat and fruits the safest, aversions to meat should be more common in pregnant women than fruit aversions. An analysis of the food preferences of 12,000 pregnant women showed that this was true. Pregnant women are ten times more likely to develop aversions to meat than to fruits.

WHY SHEEP BRAINS ARE DELICIOUS IN BEIRUT AND REPULSIVE IN BOSTON

Given our evolutionary heritage, it seems clear that humans are naturally attracted to meat. But while meat-eating seems to occur in every human society, when it comes to the amount and types of animal flesh we consume, culture rules. Worldwide, on average, humans eat about ninety pounds of meat a year. But the variation among nations is stunning. In terms of quantity, Americans and Australians vie for first place, at nearly 240 pounds of meat and fish per capita. The average person living in India, on the other hand, consumes a little over ten pounds of meat in a year. China illustrates a principle called Bennett's law—the desire to eat animals goes up as national wealth increases. While per capita meat consumption in China is presently about half that of the United States, with a billion and a half people, the numbers add up. As Marta Zaraska points out in her book *Meathooked: The History and Science of Our 2.5-Million-Year Obsession with Meat*, if the Chinese ate as much animal flesh as Americans, they would consume over 70% of the meat produced on earth.

Cultures, of course, differ in their preference for flesh. My favorite treatise on meat is a slim volume published by a Dane named Peter Lund Simmons in 1859, entitled *The Curiosities of Food or the Dainties and Delicacies of Different Nations Obtained from the Animal Kingdom*. Simmons chronicles the extraordinary range of animal flesh that humans choose to eat, most of which I would not touch. He describes the joys of elephant toes (pickle them in strong vinegar and cayenne pepper) and the gustatory delights offered by creatures such as porpoises, wombats, coffee rats, toads, bees, centipedes, spiders, sea slugs, and flamingos (the tongues of which are "extremely rich, much like that of the wild goat").

As an adventurous eater, I would rate myself a seven on a ten-point scale. I have consumed sheep brain (which I ate occasionally when I was a student in Beirut—fried is better than boiled), pig intestines, land snails, jellyfish, snapping turtle, sheep stomach stuffed with offal, grasshopper tacos, thymus glands, alligator, and the feed of chickens. However, I find yogurt disgusting, and sushi bland. I could not bring myself to eat cat, rat, bat, or chimpanzee. Neither would I eat *balut*—the Filipino delicacy which consists of a warm half-fledged duck embryo sipped from the shell. But despite my failures as a gourmand, all of these foods are regarded as delicacies in some parts of the world.

Why is the list of edible animals so long yet the number of creatures whose flesh we regularly eat so short? One reason is availability. In his book *Guns, Germs, and Steel,* Jared Diamond points out that while most animals are edible, few are good prospects for domestication and large-scale agricultural production. For example, only fourteen of the world's 148 large terrestrial mammals have been domesticated. Your options in meat depend largely on where you live. The meat counter at the supermarket where I shop only carries the standards—beef, pork, and chicken, with a few packages of lamb and a couple of kinds of fish thrown in. For the brave, there is liver. But, if you are reading this in Barcelona, you can trot over to la Boqueria, the cavernous central market on La Rambla. About halfway down the last aisle on the right, you will find the viscera monger's stall. Get there early and it will be piled high with shimmering innards—stomachs, brains, tongues, intestines, lungs, hearts, kidneys, even a couple of skinned sheep heads.

Lack of availability, however, is only one reason why people avoid eating certain types of meat. Personal experience also comes into play. Like rats, humans have evolved a special ability to associate the taste of a food with nausea and vomiting. This was discovered by the psychologist Martin Seligman who immediately developed an intense aversion to one of his favorite foods, steak with béarnaise sauce, after he came down with a virus and threw up all night after eating it on his birthday. Not surprisingly, learned aversions to meat are three times more common than aversions to vegetables and six times more common than fruit aversions.

The most important influence on whether we find a food delicious or disgusting, however, is culture. Daniel Fessler, an evolutionary anthropologist at UCLA, has studied food taboos across human societies. Because meat is dangerous, Fessler reasoned that meats should be more frequently tabooed than plant-based foods in culture. He and Carlos Navarrete collected information on forbidden foods in seventy-eight cultures. They found that perfectly edible meats were six times more likely to be tabooed than vegetables, fruits, or grains.

Why should it be easier to taboo meat than plant foods? Anthropologists love questions like this. As is often the case, there is a lot of speculation and little convincing evidence. Some food anthropologists argue that taboos exist for good reasons. Pork, for example, is tabooed by both Muslims and Jews. Those in the food functionalist camp believe that the taboo on pork serves to protect humans from trichinosis. Another functionalist view is that the prohibition on eating pigs was adaptive because swine compete with humans for the same types of foods. Similarly, the anthropologist Marvin Harris argued that the veneration of cattle among Hindus in India is adaptive because cattle are more useful for plowing fields and producing milk and fuel (dried dung) than as a source of protein.

In recent years, functionalist explanations of meat prohibitions have not fared well. They do not, for instance, explain the geographic differences in meat taboos. Why, for example, are cattle not venerated in Pakistan where, as in India, they till the soil and produce milk and fuel? Nor do they explain ecologically paradoxical taboos such as the prohibition against eating fish among desert dwellers such as the Navajo Indians of the American Southwest or the pastoral Masai in Africa. An alternative theory is that meat taboos are a result of the quirks of the human mind. I suspect that most meat taboos are simply the result of arbitrary cultural traditions and they have no explanation other than the human tendency to copy each other.

If I am correct, under the right circumstances, our feelings about the edibility of a species can change quickly, just like popular baby names and dog breeds. This was indeed the case with a taboo against eating buffalo meat among the Tharu people of Nepal. For several years, the anthropologist Christian McDonaugh lived in a Tharu village. During this time, McDonaugh regularly ate pork, goat, fish, chicken, and even rats with the villagers—but never buffalo. Buffalo and other animals were slaughtered during religious rituals. But unlike the carcasses of chicken, pigs, and goats that were eaten after the rituals, dead buffalo were dragged off and discarded. Twelve years later, McDonaugh returned to the village. He was shocked when he was offered a snack of buffalo meat toward the end of a long afternoon of cards and beer drinking. The Tharu, it seems, had changed their attitudes about buffalo flesh. McDonaugh attributes the rapid erosion of the buffalo taboo to several factors. First, the price of other meats had gone up, making buffalo a bargain. Second, the caste system was eroding. The population of the valley was becoming more diverse and the Tharu were exposed to people who did eat buffalo meat. Finally, the region was becoming more democratic and the Tharu felt freer to express their political opinions and aspirations. For the first time, they felt they could eat whatever they wanted to.

DOGMEAT COOKIES, DOGMEAT STEW: A CASE STUDY OF DIETARY TABOOS

When a culture taboos a type of meat, even the idea of eating it becomes revolting. For most Americans, the idea of consuming dogmeat is particularly repulsive. The archeological evidence, however, indicates that humans have been eating dogs for thousands of years. In many parts of the world, people historically treated dogs as walking larders to be filled up during flush times by feeding them excess food and then harvesting them when protein was in short supply. The Aztecs developed a hairless breed expressly for eating, and dogmeat was a staple among many North American Indian tribes. Though it was outlawed in 1998, dogs are still on the menu in parts of the Philippines. The anthropologist Frederick Simmons found dog eating was historically widespread in Africa, and villages in which dogmeat was relished and cultures where it was reviled were sometimes right next to each other. You would not want to be a dog in the Congo Basin where dogs are slowly beaten to death in order to tenderize their flesh.

Dogmeat is still popular in parts of Asia where approximately thirty million dogs are consumed each year. The anthrozoologist Anthony Podberscek has studied the Asian trade in dog products. He reported that the Chinese eat more dogs than anyone else, and puppy hams are the preferred cut. Compared to beef, dogmeat is a bargain. In Guangxi province, dogmeat retailed for about three dollars a pound in 2020 compared to six dollars a pound for beef. In the 1990s, dog farmers (or should they be called ranchers?) decided to produce a faster-growing animal with better meat. After experimenting with Great Danes, Newfoundlands, and Tibetan mastiffs, they chose Saint Bernards as the best breed stock because of their good temperament and ability to pump out large litters of fast-growing puppies. But because Saint Bernard flesh tends to be bland, they were often crossed with local breeds to produce better tasting meat. Meat-dog puppies are harvested at six months of age when they are still tender and juicy.

South Koreans also have a long tradition of eating dog. In Korea, as in China, dog flesh is believed to have medicinal properties. Unlike China, where dogmeat is usually eaten in the winter, Koreans consider dog a summertime fare. Despite its status as a traditional food, dogmeat has become controversial in Asia. The per capita consumption of dogmeat in South Korea is only about eight ounces a year, but with a population of fifty million, the numbers add up. The National Dog Meat Restaurant Association was organized in 2002 to promote the consumption of dogmeat and related products. These included dogmeat bread, dogmeat cookies, dogmeat mayonnaise, dogmeat ketchup, dogmeat vinegar, and dogmeat hamburger. You can also buy packs of "digested dogmeat." (I am not sure what this is.) A medicinal tonic called *gaesoju* that is said to be good for rheumatism is also produced from dogs.

While South Koreans still eat a million dogs a year, an increasing number of South Koreans are bringing dogs into their homes as pets. Cute little breeds—Maltese, shih tzus, and Yorkshire terriers—are particularly popular. As a result, South Koreans are increasingly ambivalent about eating dogs, and a 2020 poll by Humane Society International found that nearly 60% of South Koreans supported a ban on dogmeat. The shine on dogmeat is also dimming in China, where, in 2020, the government issued a directive classifying dogs as "companion animals" and that, in principle, would prohibit the sale of products made from dog flesh.

Taboos on eating dogs stem from two conflicting facets of our attitudes toward other species: We don't eat animals we despise, and we don't eat animals we dote on. The never-eat-despicable-animals principle explains why dog eating is uncommon in India and most of the Middle East. In classical Hinduism, dogs were considered the outcasts of the animal world. They were despised because they were said to have sex with their own family members and eat vomit, feces, and corpses. Dogs were likened to the people on the lowest rung of the caste system. Brahmins thought that a dog could pollute food just by looking at it. Most interpretations of Islamic law also regard dogs as unclean. Muslims, for example, are not supposed to pray immediately after being touched by a dog. Hindus and Muslims don't eat dogs for the same reasons that Americans don't eat rats—they are vermin.

Americans and Europeans don't consume dog flesh for exactly the opposite reason. Dogs in American households are not animals—they are family members. And because family members are people, eating a dog is tantamount to cannibalism.

What about cultures in which dogs can be either family or food? These societies typically have mechanisms that resolve the potential conflation of categories. The preferred breed of meat dog in South Korea are *nureongi*, midsized yellow animals that look disconcertingly like Old Yeller. *Nureongi* are not considered pets. In markets in which both pet dogs and *nureongi* are sold, the pets are physically separated from the meat dogs and housed in different colored cages. The Oglala Indians of the Pine Ridge Reservation in South Dakota consume dog stew as part of religious rituals, and yet they also keep pet dogs in their homes. The fate of each puppy in a litter is decided soon after it is born. The pets are named; their siblings destined for the stew pot are not.

THE TYRANNY OF THE FORK

So while it tastes good, meat is dangerous and easy to taboo. Another factor that clouds the human-meat relationship is associated with the guilt that comes from what the philosopher Tom Regan called "the tyranny of the fork"—taking the life of another creature. In tribal societies, ceremonies in which hunters atone for killing animals are nearly universal. Most Americans, however, avoid meat-related guilt simply by not thinking about where their dinner comes from. I successfully avoided the moral consequences of my diet until, at the age of thirty-six, I found myself, skinning-knife in hand, hacking away at the steaming body of a 1,300-pound steer.

At the time, we were living on the campus of Warren Wilson College near Asheville. The college maintained a farm that included a herd of beef cattle, about thirty of which were slaughtered each year. These animals led an idyllic pastoral life and had the painless death that even animal liberation philosopher Peter Singer would probably not object to. They were never crowded into feedlots or jammed into tractor-trailers or subjected to the trauma of an industrial slaughterhouse. No. Warren Wilson cows were doted on from birth to death by the earnest back-to-the-land students who worked on the college farm crew. And, the morning it was slaughtered, each steer was persuaded with a handful of sweetgrass to walk into a small abattoir where it was shot in the head before it could say moo.

Some of the students on the farm crew knew that I studied the psychology of human-animal interactions, and one afternoon they suggested that I help them slaughter cattle the next day. After hemming and hawing a bit, I reluctantly agreed. I didn't sleep much that night. I showed up at the small abattoir on the campus at seven the next morning, and an hour later I was up to my elbows in bovine entrails. I spent the next two days helping convert big animals into packages of chilled meat.

Here's how the first steer went down. One of my students, Sandy McGee, led the animal into the kill room and tied its halter to a ring on the floor. The farm manager walked into the room with a .22, shot the steer in the head, and the students went to work. Without being told, they knew what to do. One of them slit the steer's throat, bleeding out the carcass, while another sawed off the animal's head and hooves. They attached a chain to the steer's legs and hoisted the carcass vertically toward the ceiling. Out of nowhere, a wheelbarrow appeared—I did not have a clue what for. Then, using a six-inch skinning knife, Sandy made a quick vertical incision, opening the animal from rib cage to anus. Gallons of viscera plopped overflowing into the wheelbarrow. The USDA inspector examined the heart, liver, and kidneys, and stamped the carcass fit to eat.

You sometimes hear that if we had to kill our own meat, everyone would be a vegetarian. The students on the farm crew offered an opportunity to test this theory. Most of them were raised in middle-class suburbs and had never been around a cow or hog until they came to college. To assess the validity of the "slaughter leads to meat rejection" hypothesis, Sandy and I distributed questionnaires to the three dozen students who regularly participated in killing and butchering cattle and hogs. I also interviewed most of them.

Our results refuted the idea that slaughtering in and of itself creates vegetarians: None of the student slaughterers had given up meat. But their responses to slaughtering were complicated. While nearly all of them said they found the killing and butchering process an interesting and valuable experience, most admitted that they sometimes felt nauseous during or after butchering a cow. Half of the farm crew said they sometimes felt guilty and sometimes avoided eating meat for a day or two after they had slaughtered a cow or pig, but most of them felt they had benefited from the experience of slaughtering animals. Some of their reasons were disappointingly mundane. For instance, several told me they enjoyed learning about different cuts of meat and felt it would make them better shoppers. A couple of pre-veterinary majors said that butchering had helped them learn anatomy. But for others, the experience had a deeper meaning. It was an exercise in values clarification. They had learned where meat comes from. That it is dead.

Food psychologist Paul Rozin argues that animals and death are intimately connected in the human psyche. Rozin believes that people find many animal products, including their flesh, disgusting because animals are an uncomfortable reminder of our mortality. He wrote, "Humans must eat, excrete, and have sex, just like animals. Each culture prescribes the proper way to perform these actions—by, for example, placing most animals off-limits as potential foods, and all animals and first-degree relatives off-limits as potential sexual partners. Furthermore, we humans are like animals in having frail body envelopes that, when breached, reveal blood and soft viscera that display our commonalities with animals. Human bodies, like animal bodies, die."

Do people actually find animal flesh disgusting? Increasingly, the answer is yes. For instance, researchers found that the redder and more "animalized" a cut of meat is—the more it resembles a carcass—the more it turns the average consumer off. These findings pose a conundrum for the meat industry. Usually, shoppers are drawn to foods that appear fresh, juicy, and natural. But in the case of meat, fresh and juicy is seen as gross, particularly to women. The investigators recommended that the meat industry develop products that look less like, uh . . . meat—small, ready-to-cook cuts marinated to mask the smell and color of flesh—less disgusting meat.

Chicken producers caught on to this a long time ago. In the 1960s, almost all the chickens sold in the United States were purchased as an intact carcass with heart, liver, and gizzard tucked neatly into the body cavity. You had to hack them up yourself. Today, fewer than 10% of chickens sold in supermarkets bear any resemblance to the body of an animal. The fastest-growing segment of the retail chicken industry is officially referred to as "further processed"—translucent, boneless pieces of flesh that look like they were grown in a Petri dish and labeled something like "tenders" or, my favorite oxymoron, "chicken fingers."

IF MEAT IS SO DISGUSTING, WHY ARE THERE SO FEW VEGETARIANS?

The process by which initially neutral preferences come to be regarded as immoral is called "moralization." Attitudes toward slavery have been moralized as have, more recently, cigarette smoking. You would think meat would be easy to moralize. The floor of my office is piled with books telling me why I should not eat animals. The case against meat boils down to four claims that are hard to dispute. First, to eat an animal, you have to take its life. Second, the conditions under which nearly all meat animals are raised, transported, and slaughtered involve unspeakable suffering for the animals and horrible conditions for the people who do the dirty work. Third, the conversion of plants to meat is inefficient and environmentally destructive. Fourth, eating animals causes obesity, cancer, and heart disease. Add the *yuck* factor to the moral and health arguments against meat, and you would think it would be easy to convince people not to eat flesh.

But you would be wrong. In fact, the inability to convince people to give up meat is the most glaring failure of the animal protection movement. In 1975, when Peter Singer jump-started the animal rights movement with the publication of *Animal Liberation: A New Ethics for Our Treatment of Animals*, 98% of Americans ate meat. In 2019, the percentage of Americans who ate meat had barely nudged down to 96%. In comparison, the campaign to moralize not smoking was much more successful. During the same period, the percentage of American adults who smoked dropped from 47% to 14%.

These days, it is fashionable, especially among the young, to claim "I don't eat meat." In a survey, 30% of college students said that having a vegan option at every meal was important to them, and sales of "faux meat" in the United States are growing at a rate of 35% a year. And three out of four respondents in a 2018 national survey agreed that "knowing the animal did not suffer when it was raised on the farm" was an important factor in their purchasing decisions. There is, however, little evidence to support the view that a wave of vegetarianism is sweeping across America. Some of the best estimates of the number of vegetarians in the United States come from surveys conducted regularly since 1994 by the Vegetarian Resource Group. These polls consistently show that between 95% and 98% of American adults eat flesh.

Ironically, as our collective concern for the welfare of animals has increased, so has our desire to eat them. In the 1970s, the average American ate 176 pounds of meat and fish a year. In 2020, we were up to almost 240 pounds a year. The change in the number of meat animals slaughtered annually is even more astounding. Over the last forty years, the number of creatures killed for our dining pleasure jumped from three billion to ten billion; from fifty-six animals a year for a family of four to 132 animals.

Why has the animal protection movement had so little effect on our diet? Ironically, the efforts by animal protectionists to improve the well-being of farm animals have made the consumption of flesh more, rather than less, morally palatable. Socially conscious consumers can now purchase meat touted as hormone-free, antibiotic-free, cruelty-free, and free-range. In other words, guilt-free. The supermarkets in my town are stocked with flesh which has been given the moral thumbs up by the American Humane Association. I can buy chickens that, I am told, led the good life: "100% all-natural," "fed an all-vegetable diet!" under "low-stress growing practices" with "no debeaking," and "tunnel ventilation for fresh air," and "multiple feed bins to ensure fresh feed." The fast-food chains have jumped on the animal welfare bandwagon. McDonald's, Wendy's, and the parent companies of KFC and Hardee's have established high-powered animal welfare advisory boards and adopted animal care and slaughter standards for their suppliers. And I was the first in line when my local Burger King rolled out their meatless Impossible Burger.

The reason for the huge jump in the number of animals killed in American slaughterhouses was the shift from eating mammals to eating birds. For many years, one of my guilty pleasures was diving into a BK Whopper with Cheese. I loved the gooey mayonnaise sauce, iceberg lettuce, juicy fats, and charbroiled flavor that I figured came from a New Jersey chemical plant. My Whopper habit evaporated immediately after I read Eric Schlosser's book *Fast Food Nation: The Dark Side of the All-American Meal*. It did not take much to convince me that Coke was as addictive as heroin, that McDonald's executives had conspired to hold down the minimum wage, and that the fast-food industry had done nearly as much harm to Americans as Purdue Pharma. A Whopper did not taste nearly as good to me after I found out that a ground beef patty contains bits and pieces of four hundred cows, any one of which could be sick, and all of which spent the last weeks of their lives standing knee-deep in manure.

I was not alone in reducing my intake of steak and burgers. Per capita beef consumption in the United States began a twenty-year slide in the 1970s when the Food and Drug Administration told us to cut down on saturated fats. The drop in our enthusiasm for beef, however, was more than compensated by an extraordinary increase in our desire to eat chicken. The number of cattle slaughtered each year in the United States dropped 20% between 1975 and 2009, while the number of chickens killed increased by 200%. The watershed year was 1990 when, for the first time in history, Americans ate more chicken than beef. When Herbert Hoover ran for president on a campaign of "a chicken in every pot," the average American ate half a pound of chicken a year. Today, the figure is one hundred pounds per person.

There were several reasons for our shift from eating cows to eating birds, and they have nothing to do with a growing concern for animal welfare. Advances in poultry science and the vertical integration of the chicken industry since the end of World War II have made chicken a better deal than beef. In 1960, a pound of chicken cost half as much as a pound of beef; now it is only a quarter of the price of beef. Beef became linked with obesity, cardiovascular disease, and cancer. While some early claims against the hazards of red meat were based on shoddy science, epidemiological studies have confirmed that eating cow is bad for your health. A multisite study of half a million people found that the individuals who consumed a lot of red and processed meat were more likely to die from cancer and cardiovascular disorders than people who ate less red meat. The authors of the report estimated that the death rates of Americans would drop 11% in men and 16% in women if they ate less red meat.

My nutritionist friend Cathy's solution for these health issues is to tell her clients to eat only animals that swim or fly. But from an animal welfare perspective, the movement away from beef has been a disaster. The average steer weighs about 1,100 pounds at slaughter, 62% of which is useable meat. A Cobb 500 broiler, in contrast, yields about three pounds of meat. This means that you need to kill 200 chickens to get one cow's worth of meat. Further, cattle enjoy longer and more pleasant lives than factory-farmed chickens. While a Cobb 500 will breathe ammonia fumes 24/7 and never see the sky, the average steer will spend a year and a half munching grass in a sunny pasture before it is hauled off to a feedlot for "finishing." A McDonald's Caesar salad with grilled chicken might be better for your health, but in terms of animal suffering, the moral scales tip toward a Big Mac.

Of course, by this logic, the food of choice for the 40% of animal activists who sometimes eat meat would be a whale. There are 70,000 chickens worth of flesh in a single 100-ton blue whale. Ingrid Newkirk, co-founder and president of PETA, agrees. In 2001, to the dismay of some PETA supporters, the organization launched a campaign urging people to eat whales. Here's how Newkirk explained PETA's logic to me in an email:

> We started the "Eat the Whales" campaign to draw attention to the fact that the bigger the animal, the more meals that could be obtained from just one animal's suffering and death. In the case

of whales, there is the added benefit that the animals lived free, didn't have their ears notched or their tails cut off, weren't castrated or debeaked, were never crammed into a cage that rubbed their flesh raw, never stuffed into a transport crate in all weather extremes, and so on. So, yes, to spare the greatest number of animals from suffering, if you can't (or won't) shake off that meat addiction, if you will not abandon that fleeting taste of meat for the sake of compassion and decency, your own health, and the environment, then it is better to eat body parts from the largest animal you can get your hands on.

Makes sense to me. But my daughter Betsy, who lived for a year in rural Japan, says whale meat is tough, stringy, and gross.

ARE FISH MADE OUT OF MEAT?

While the campaign to moralize meat in the United States has not been a big success, between eight and thirteen million Americans identify as vegetarians. Michele, one of my graduate students, was one of them. While chewing on a piece of seared ahi tuna, she nonchalantly told me that she does not eat meat. She is not unusual; most "vegetarians" in the United States eat animal flesh, often fish.

Not Che Green. Che is the founder of Faunalytics, an animal welfare organization that conducts high-quality studies of shifting public attitudes about the treatment of animals. Like most animal protectionists, Che had a soft spot for other creatures when he was a little kid. He ate meat as a child, though he much preferred dishes that did not remind him that he was eating an animal. His attitudes toward meat changed in high school when he landed a summer job in an Alaskan cannery. His task was to feed big fish into a processing machine where they were spit out a few seconds later as canned salmon. He made it through the summer, but the carnage got to him. Within two months he had become a vegetarian and two years later, a vegan.

Che and the researchers at Faunalytics have conducted surveys of vegetarians and vegans in the United States for years. Their studies illustrate a fundamental principle of human psychology—what people say is often different from what they do. For example, a USDA survey of over 15,000 Americans found that 48% of people who claimed they were vegetarians admitted that they had eaten red meat, poultry, or seafood within the previous twenty-four hours. Other studies obtained similar results. The paradox that more than half of "vegetarians" eat some form of animal flesh daily raises an interesting question—what exactly counts as meat?

Like his father, my grandson Ryland loves barbecue and burgers. His mother, however, is a longtime vegetarian. Ever the psychologist, I was interested in how being raised by a vegetarian mom and an omnivore dad affected his perceptions of meat. So when he was five, I tried to convince him that people, like the creatures we eat, are made out of meat. After all, for tens of thousands of years, our ancestors were prey as well as predators. Even today, roughly 1,500 people are eaten by lions, tigers, or crocodiles every year. Ryland, however, was not having it. So, together, we made a list of animals he believed were and were not made out of meat. In the "meat" column were cows, birds, pigs, fish, and, for some reason, "monkeys in Africa." He put elephants, chimpanzees, Big Bird, and people in the "not-meat" column.

I thought Ryland's meat taxonomy was pretty wacky until I read a 2018 study on the diets of vegetarians. The lead author was a UCLA graduate student named Daniel Rosenfeld whose résumé contains an impressive list of publications on the psychology of eating. Daniel and his colleagues were interested in comparing the beliefs of pescatarians (people who eat fish but no other type of flesh) with strict vegetarians. Thirty-seven percent of people in their study who ate fish also claimed to be either vegetarian or vegan. The biggest surprise was in their culinary taxonomies. It turned out that the pescatarians were not much different than my five-year-old grandson. Nearly half of them denied that fish were made out of meat. The researchers wrote, "Viewing fish as distinct from other meat may be a strategy for reducing the cognitive dissonance and threats to one's moral self-concept one might feel from viewing oneself as a meat-eater."

Go figure.

WHY PEOPLE GIVE UP MEAT

People give up meat for a variety of reasons. Health concerns are the primary motivator for most vegetarians, with moral and environmental concerns coming in a close second. Sometimes motivations for going vegetarian evolve. Che Green initially gave up meat because of his visceral disgust associated with working at the canning factory. His subsequent veganism was a consequence of his growing involvement in the animal rights movement. My friend Pete's road to vegetarianism was different. His parents were Seventh-Day Adventists who did not eat meat for religious reasons. But these days, Pete is primarily motivated by the health advantages of a plant-based diet. Pete's commitment to vegetarianism has little to do with concern with the rights or suffering of other creatures. He does use a large Havahart live trap to humanely capture the animals that raid his garden. But then he shoots them in the head.

Pete lives on a mini-farm north of Asheville, North Carolina, where he raises much of the food his family consumes. Five years ago, he grew tired of sharing his corn, squash, peas, beans, and blueberries with the growing population of animals who also relished fresh vegetables. He purchased a couple of live traps, and he would catch the animal invaders and release them a couple of miles away. When that did not work, he bought a gun. Last year he killed two raccoons, a couple of turkeys and a possum. Over the past five years, he has shot fifty animals to protect his veggies. Pete takes no pleasure in the killing, and he is constantly trying to improve the fencing and netting around his garden so he will not have to kill the animals raiding his veggies. But at this point, the raccoons still manage to wreak havoc on his corn, and he remains a vegetarian hunter.

As illustrated by Che and Pete, vegetarians do not necessarily think alike about the morality of eating flesh. Paul Rozin and his colleagues found moral-origin vegetarians are more disgusted by meat than health-origin vegetarians, and they are more upset at the prospect of chewing and swallowing flesh. Unlike health-motivated vegetarians, moral vegetarians tend to see meat as contaminating and meat-eaters as aggressive. They also tend to have more extensive rationales for eschewing flesh, and they reject more animal products than health-origin vegetarians do. They are also more likely to ascribe emotions to animals, to own pets, and feed their pets vegetarian diets. Further, ethical vegetarians find meat more disgusting, and they are less likely to go back to eating meat. In other words, ethical vegetarians moralize meat more than health vegetarians.

Demographics also play a role. In the United States, vegetarians are more likely than the average American to be female, white, and college-educated. Vegetarians and semi-vegetarians tend to score higher on personality measures of anxiety and openness to new experiences and, as you would expect, they have more pro-animal welfare attitudes. They also tend to be liberal when it comes to issues like gun control, the legalization of marijuana, redistribution of wealth, and universal medical care. They usually vote Democratic. Most vegetarians give up red meat first, and then expand their list of rejected foods to chicken and fish, and, in the case of vegans, eggs and dairy products. The motivations of vegetarians often evolve over the years. A person who initially gives up meat for health reasons may subsequently internalize the moral arguments against eating animals.

THE CONNECTION BETWEEN VEGETARIANISM AND MENTAL HEALTH ISSUES?

In their political views, vegetarians tend to differ from meat-eaters. But does this also apply to their mental health? The results are mixed. For example, some studies have linked vegetarianism to eating disorders. In recent years, researchers have begun to examine relationships between diet and other forms of mental illness, particularly depression. Many of my friends have given up meat, and most of them are happy. My colleague Mickey is a bundle of energy and she laughs a lot. Arguing over beers with my vegan pal Dave about whether vegetarians live longer is always fun. And it was a hoot working with Shelley Galvin back when we were combing through issues of supermarket tabloids for weird stories about human-animal interactions.

Therefore, I was surprised when I came across an article that reported vegetarians were more likely to be depressed than meat-eaters. Intrigued, I quickly located nearly a dozen research papers published between 2007 and 2020 which compared rates of depression in meat-eaters and meat-avoiders. The good news is that in all the studies, most of the vegetarians did not have mental health issues. The bad news is most of the studies did report that depression was more common among vegetarians than meat-eaters. German researchers examined the frequency of mental health problems among a representative sample of over 4,000 vegetarians and non-vegetarians. Even though the groups were matched on a slew of demographic and socioeconomic variables, the vegetarians were more likely to have suffered from depressive disorders in the previous month, the previous year, and over their lifetimes. Austrian investigators conducted a particularly elegant study involving 330 vegetarians, 330 people who consumed a lot of meat, 330 omnivores who ate less meat, and 330 people who consumed a little meat but ate mostly fruits and veggies. The participants were carefully matched for sex, age, and socioeconomic status. The researchers found that vegetarians in the study were twice as likely as the other groups to suffer from mental illnesses including depression. Catherine Forestell reported that vegetarian and semi-vegetarian college students scored higher than student omnivores on a standardized depression scale. And a team of European researchers reported that the frequency of seasonal affective disorder was four times higher among Finnish vegetarians than omnivores.

At least from a research point of view, it would be great if the evidence connecting vegetarian and vegan diets to depression were a slam dunk. But, alas, this is not the case. A French study concluded that vegetarians were no more likely to suffer from depression than meat-eaters. (They were, however, less happy than the omnivores.) Researchers from Benedictine University found no differences in depression scores among vegans, vegetarians, and omnivores in their study. And investigators from Arizona State University reported that Seventh-Day Adventists who were vegetarians had lower depression, anxiety, and stress scores than meat-eating Adventists.

So, some studies have found vegetarians and vegans are more likely than meat-eaters to suffer from depression and some have not. Conflicting results are common in the medical and behavioral sciences. Investigators use a technique called systematic review to help make sense of inconsistent results. In 2020, a team led by Urska Dobersek of the University of Southern Indiana took on the growing body of research on vegetarianism and mental health. Eighteen studies met their selection criteria. They carefully examined the results of each study and the quality of methods. Their conclusions were clear.

> Based on this systematic review comprising 160,257 participants from varied geographic regions, including Europe, Asia, North America, and Oceania, aged eleven to ninety-six years, there is clear evidence that meat-abstention is associated with higher rates of depression, anxiety, and self-harm.

A similar study-of-studies published in 2020 in the journal *Nutrition Reviews* examined the results of fourteen published research articles involving nearly 18,000 individuals. The authors concluded that the rates of depression were twice as high in nonmeat-eaters as omnivores.

There seems to be a curious link between depression and going veg. But we have to be careful about "link-think." As I pointed out earlier, childhood animal cruelty has been linked to adult violence, yet most kids who abuse animals grow up to be perfectly normal adults and most pit bulls never bite anyone. Likewise, only a small fraction of people who suffer from depression are vegetarians and most vegetarians are not depressed. However, the findings of multiple studies involving thousands of subjects in different countries suggest that the connection between vegetarianism and depression is not just a statistical fluke.

What could be going on? The association between giving up meat and depression yet again raises the pesky causal arrow problem. It is possible, I suppose, that vegetarian diets could produce physiological changes in our bodies that cause depression. But I am skeptical of this explanation. And in their systematic review, Dobersek and her colleagues found no evidence to support the idea that avoiding meat *causes* poor mental health. I think it is more likely that some traits may predispose some people to both depression and the decision to become a vegetarian. Women, for example, are twice as likely as men to suffer from depression, and there are also more female than male vegetarians. Personality traits might also predispose people to both vegetarianism and depression. Personality researchers have found that depressed people tend to score high on the Big Five trait of anxiety as do vegetarians.

Depression in vegetarians could also result from social isolation. Across all cultures, much human social life revolves around the sharing of food and drink. It's safe to say that most moral vegetarians would not feel comfortable sitting at a dinner table with family or friends where the main course was a hunk of rare rib roast. An older woman in a class I was teaching recently told me that her friends had stopped asking her over to their houses after she became a vegan. "They don't know how to cook for me," she said. Her experience is consistent with the findings of the social psychologists Cara MacInnis and Gordon Hodson. Twenty-five percent of vegans they studied reported their friends did not hang out with them much after they had come out of the vegetarian closet.

The commitment to not exploiting animals could also be a source of depression in ethical vegetarians. Sometimes inadvertent or minor transgressions can be upsetting. For example, a woman posted a cry for help to a vegan Facebook group:

> Guys, I need some emotional support from the group. I just ate lima beans, and I realized after eating a few bites that they had pig fat in them. I feel so gross and don't know what to do.

The animal activist/vegan/neuroscientist Lori Marino concurs that the commitment to animal rights can itself lead to depression. As she wrote to me in an email:

> Vegetarians and vegans are more aware of the cruelties of the world, and this is more depressing than living in a state of ignorant bliss.

WHY DO MOST VEGETARIANS RETURN TO EATING MEAT?

My pals Joanie and Phil have been married for over twenty years and they take animal issues seriously. They oppose all animal research, they don't wear leather, and they don't consume any animal products. We occasionally get together for lunch at one of Asheville's vegetarian or vegan restaurants. As we finished our meal a couple of months ago, Joanie quietly said she had something to tell me. She wanted me to know that after fifteen years of strict veganism she had started eating meat. For months, Joanie had been chronically tired and had gained weight. A series of tests indicated her blood chemistry was screwed up. Her doctor suspected her problems were related to eating excess carbs. She warned Joanie that her health would probably continue to go downhill unless she began to consume some animal protein. After giving it a lot of thought, Joanie decided to eat a little chicken a couple of times a week. It worked. She told me she was feeling a lot better.

Joanie is not unusual. Indeed, a survey of over 11,000 American adults conducted by Faunalytics found that reverting back to meat is more the rule than the exception among vegetarians and vegans. Eighty-six percent of the vegetarians and 70% of vegans in their study had returned to being omnivory. Perhaps because I was raised a Southern Baptist, I have always been fascinated by backsliders—people who have seen the light but then have a change of heart. I ran the idea of studying ex-vegetarians by Morgan Childers, an honors student who came into my office one afternoon looking for a research project. We designed an online survey and Morgan recruited participants by sending announcements of the study through social media interest groups.

Within a couple of weeks, seventy-seven former vegetarians had completed our questionnaire. On average, they had been vegetarians for nearly ten years before they resumed eating meat. In his book *The Face on Your Plate: The Truth About Food*, Jeffrey Moussaieff Masson extolled the health benefits of avoiding animal-based foods. He wrote, "Vegetarian for most of my life, I have never really experienced illness. Now at sixty-eight, several years a vegan, I find that I have never been healthier: I weigh less than I did at thirty; I am stronger than I was at forty; I have fewer colds or minor illnesses than at fifty; and, in my entire life, I have experienced no major illness of any kind."

He is lucky. Poor health was the most common reason vegetarians in our study had resumed eating animals. Recall that Staci reverted to being an omnivore because she always felt sick. Many of the ex-vegetarians that took our survey said the same thing. One wrote, "I was very weak and sickly. I felt horrible even though I ate a good variety of foods like PETA said to." Another said, "I was very ill despite having regular iron injections and vitamin supplements. My doctor recommended that I eat some form of meat as I was not getting any better. I thought it would be hypocritical of me to just eat chicken or fish as they are just as much an animal as a cow or pig. So I went from no meat to all meat." The most succinct response was from a person who wrote, "I will take a dead cow over anemia any time."

But there are other reasons that vegetarians return to meat. Many of our participants simply grew tired of the hassle of vegetarian or veganism—they could not find good-quality organic vegetables locally or at a price they could afford, they did not have the time to prepare vegetarian meals, or they simply grew tired of the lifestyle. In describing the dietary difficulties he faced, the philosopher Gary Steiner wrote, "You just haven't lived until you've tried to function as a strict vegan in a meat-crazed society. What were once the most straightforward activities become a constant ordeal."

Some vegetarians suddenly develop cravings for animal flesh. A woman named Charlene wrote to a Facebook vegetarian group recently, "I'm forty-years-old and four years a vegetarian! I lately have been craving seafood like crazy! I ate shrimp yesterday for the first time in my life. The craving is ridiculous! I need more vitamin D and zinc, I guess. I should add, I've never liked seafood much at all." About a third of the former vegetarians in the Faunalytics study indicated that missing the taste of red meat, poultry, or seafood was a factor in reverting back to meat. Some of the participants in our study talked about protein cravings or how the smell of sizzling bacon drove them crazy. One wrote, "I just felt hungry and that hunger would not be satisfied unless I ate some meat." Another was succinct: "Starving college student + First night back home with folks + Fifty or so blazin' Buffalo wings waiting in the kitchen = Surrender."

MEAT AS THE BATTLEGROUND
BETWEEN MIND AND BODY

When he was a graduate student, the psychologist Jonathan Haidt decided industrial meat production was immoral when he read Peter Singer's book *Practical Ethics*. His new awareness of the cruelty inherent in industrial agriculture, however, did not affect his diet. He wrote:

> Since that day I have been morally opposed to all forms of factory farming. Morally opposed, but not behaviorally opposed. I love the taste of meat, and the only thing that changed after reading Singer is that I thought about my hypocrisy each time I ordered a hamburger.

My experience is similar to Haidt's. I grew up in a meat and potatoes family and often ate meat three times a day, usually foods that start with the letter "b"—bacon, baloney, beef, barbecue. No longer. More because we like the flavors than anything else, Mary Jean and I are drawn to Mediterranean-style cuisines that are supposed to be good for you— dishes that taste of tomato, lemon, and garlic; pasta and rice entrees. We do eat meat, though much less than we used to, and usually creatures that swim or fly.

I also make what are probably symbolic gestures to reduce cruelty. I get eggs from my friend Lydia who dotes on her mixed flock of Araucanas and Barred Rocks. I pay three times as much as I need to for chickens from Bell and Evans whose website claims their chickens were "allowed to bask in the warm light of the sun." And my occasional flatiron steak comes from a Niman Ranch steer that I am told was "humanely raised on sustainable U.S. family farms and ranches." I know, however, that according to *Consumer Reports*, terms like "natural" and "cruelty free" are marketing ploys that usually mean nothing.

Meat inhabits the psychological territory that Al Pacino called the "no-man's land in the battle between mind and body." The most natural of human interactions with animals is our desire to eat them. Meat hunger is metaphorically "in our genes" just like it is in chimpanzee genes. While people like Jonathan Haidt and myself cave when it comes to matters of the flesh, humans are the only species with the ability to look into the eyes of a chicken and decide it would be wrong to eat it.

The primatologist Marc Hauser once remarked that the cognitive chasm between humans and chimpanzees is greater than the gap between an ape and a worm. Nowhere is the difference between humans and other animals more apparent than in matters of food. Chimps can recognize themselves in mirrors, make tools, coordinate group hunts, use symbols to communicate, and establish political alliances. But no chimpanzee has ever shown the slightest sign of remorse when ripping a tasty arm off a screaming colobus monkey.

RAW STEAK DINNER: YUM OR YUCK?

A month after I interviewed Staci about her transformation from a vegetarian to a raw meat-eater, I got an email from her.

> Hal, could you and Mary Jean come over for dinner Sunday? We're having steak.

Ah . . . "Sure, Staci. What kind of wine goes with raw meat?"

A week later, I was having second thoughts, having been lectured by my son Adam (an emergency room nurse) and his wife, Alendia (a physician), about the perils of uncooked flesh. But on Sunday afternoon, we drove over the mountain. Staci gave us a tour of the farm which was in full summer blossom. Two adolescent pigs ran over to us, oinking enthusiastically; they seemed genuinely glad to meet us. Then it was time to eat. For Staci, Gregory, and me, dinner was a raw T-bone and a lovely Greek salad. (Mary Jean opted for baked chicken breast.) The steak, which came from a steer Staci and Gregory raised, was surprisingly good. Tender, tasty, moist. My reservations disappeared. I asked for seconds and even sampled a slice of raw duck Gregory offered me.

A couple of weeks later, I got an email from Staci that nicely captured the moral ambiguity of the human-meat relationship.

Hal,

We just took our pigs to the butcher this morning.

It's amazing how complex our psyches must be in order to nurture creatures every day for seven months, only to have them sent away and then come home in little freezer packages. Or sometimes to butcher them ourselves.

I think it takes bravery, don't you?

I think of all the millions of humans over time who have hunted and raised animals for food because that was the way you survived. But you need to make it right in your conscience. Maybe reverence helps. Maybe killing the creature yourself helps. It completes the cycle somehow. Taking responsibility is somehow the balm that soothes the horror.

Blessings to you and Mary Jean and to our pigs.

The Moral Status of Mice

THE ETHICS OF ANIMAL RESEARCH

> If, in evaluating a research program, the pains of a rodent
> count equally with the pains of a human, we are forced
> to conclude (1) that neither humans nor rodents possess
> rights, or (2) that rodents possess all the rights that humans
> possess. Both alternatives are absurd.
>
> —CARL COHEN, BIOETHICIST

> Here was the human body writ small.
> —ALLEGRA GOODMAN, NOVELIST

Many *Wall Street Journal* readers probably missed the brief article in the January 8, 2020, issue with the headline, "New Virus Discovered by Chinese Scientists Investigating Pneumonia Outbreak." According to the article, fifty-nine people in Wuhan, China, had suddenly come down with a fever and dry cough, and seven of them were hospitalized. Who knew that a year later, nearly one hundred million people would have contracted COVID-19, that the global death toll would be two million and climbing, and that more Americans would have died from the pandemic than were killed in World War II, the Korean War, Vietnam, 9/11, and Desert Storm combined?

Medical investigators soon discovered that COVID-19 was caused by a highly infectious novel virus named SARs-CoV-2. Thousands of scientists worldwide in government, university, and pharmaceutical company labs began a fast-track search for vaccines to immunize against the virus. According to the National Institutes of Health, studies on animals were instrumental in the early stages of the development of COVID-19 vaccines. But animal rights groups were not convinced. PETA claimed that the fast tracking of COVID-19 vaccines proved that medical progress would not be impeded if animal research was eliminated.

Dr. David DeGrazia, a noted bioethicist who works with the National Institutes of Health, takes a middle ground. He opposes painful research on animals, but he admits there are exceptions. In an interview in the journal *Science*, he said:

> If we, hypothetically, addict mice to cocaine, and then see how much of an electric shock they're willing to endure to get their fix . . . [t]hat's way too much harm to cause whatever the purpose of the study. But there might be some rare exceptions. For example, if there's a raging pandemic and the only way to test a vaccine is to have a control group that suffers for a week or more without treatment.

I wondered if COVID-19 would be one of those exceptions, so I sent him an email:

> Would you put the use of animals like monkeys and genetically modified mice in the search for vaccines and treatments for COVID-19 in the "raging pandemic rare exception" category?

His answer came the next day.

> The current pandemic is certainly the type of situation in which exceptions might be justified. But it would really have to be necessary. . . . This might be the case with using monkeys as test subjects. I am doubtful about genetically modified mouse models which tend to translate so poorly into human disorders.

It usually takes a long time to develop a vaccine—twenty-eight years for chicken pox, fifteen years for human papillomavirus, six years for polio. But with COVID-19, it only took eight months from lab bench to shots in the arm. And, despite Dr. DeGrazia's misgivings, a lot of mice were killed in the process.

■ ■ ■

My first brush with the moral complexities of animal research involved a mouse. In my second year of graduate school, I was assigned to work as a lowly assistant in the laboratory of a chemical ecologist. One of my jobs was to collect molecules from the skin surface of earthworms. The procedure involved dropping worms into 180-degree water. Two minutes later I would remove their inert bodies from the hot water and freeze little vials of *eau de worm* for later chemical analysis. I had performed this procedure several times and viewed it as just another lab chore, one that I did not enjoy, but which also caused me no moral discomfort. The worms died instantly, and, after all, they were just worms.

One morning I was asked by the lab manager to do something different. A scientist from the University of Utah who was studying the skin chemistry of desert creatures had arranged for some of his analysis to be done in our laboratory. Several days later, a box stamped "Caution: Contains Live Animals" was delivered to the lab. Inside was a virtual menagerie: a dozen crickets, a pair of eerie pale scorpions, a lizard about six inches long, a small snake, and a lovely little gray deer mouse. I was given the job of liquefying the animals.

I had plunged an occasional lobster into a pot of boiling water with only a slight moral twinge, and I did not expect to be bothered by my morning's task. I lit the Bunsen burner and started to work my way up the phylogenetic scale. Like worms, the crickets died almost immediately when they hit the near-boiling water. No problem. Next, the arthropods. In the few days they had been in our lab, I had come to like the scorpions. They had an air of menace I found fascinating. They also had more body mass than the crickets and took a little longer to die when I dropped them into the water. I began to wonder what I was doing.

The lizard was a striped juvenile of the genus *Cnemdopherous*. My stomach turned as I lifted it from its cage, and I began to sweat. My hands shook a little when I dropped it in the hot water. The lizard did not die quickly. It thrashed for maybe ten seconds before becoming still. The little snake was an elegant racer with big black eyes. More shaking hands and sweating brow, and the thrashing reptile soon was reduced to molecules swirling in solution.

Finally, the mouse. I weighed the mouse, calculated the appropriate amount of distilled water, poured it into the beaker, and turned on the heat. As the water approached 180 degrees, I realized I just could not "do" the mouse. I turned the Bunsen burner off and with a mixture of trepidation and relief, walked into the office of the lab manager, thinking that it might be my last day of graduate school. I told him that I had made extracts from most of the animals but that I just could not drop a live mouse into scalding hot water. To his credit, that was the end of it. My boss did the mouse while I waited in the next room.

I have thought about my predicament many times since. In hindsight, I am struck by the similarity between my tasks that morning and the plight of the subjects in the psychologist Stanley Milgram's infamous obedience experiments. As all introductory psychology students learn, the hapless participants in his studies were instructed to administer electrical shocks of increasing intensity to subjects in an adjacent room. The majority of people in the experiment administered shocks they thought would be extremely painful, if not lethal. Like Milgram's participants, I was confronted with a series of escalating choice points, but in my case, based on the phylogenetic scale rather than electric shock levels. The difference was that in the Milgram study the shocks were a ruse; the supposedly shocked "subjects" were actually confederates of the experimenter. In my laboratory, the animals died. When I look back on the incident, I get some satisfaction in knowing that I refused to boil a living mouse. But I wish I had quit between the cricket and the scorpion.

This event provoked me to ask myself questions that I still struggle with. What is the difference between researchers who kill mice because they are working on vaccines for new contagious viruses that could kill millions of people and the legions of good people who smash the spines of mice in their homes with snap traps or slowly poison them with d-CON? Why was it easy for me to plunge crickets into hot water, harder for me to kill the lizard, and impossible for me to boil the mouse? Was it a matter of size, phylogenetic status, nervous system development, the grisly manner of their death, or simply the fact that the mouse was really cute? Were the results of this experiment worth the death and suffering of the animals? Are they ever?

DARWIN'S MORAL LEGACY

I am not alone in my ambivalence about animal research. In the United States, public opinion on the topic is split almost evenly. According to Gallup polls, the percentage of Americans who felt that medical testing on animals was morally acceptable dropped from 65% in 2001 to 50% in 2019. In contrast, 80% of Americans approve of hunting.

Even Charles Darwin struggled with vivisection—the nineteenth-century term for invasive animal research. Darwin was fascinated by animals, and he was confronted by the problem that modern zoologists face—sometimes you wind up killing the very creatures you have dedicated your life to studying. Jim Costa, a Darwin historian, told me that as a fledgling naturalist, Darwin shot and poisoned thousands of animals, including mice, for his collections. He was even horrified by some of his own experiments. He wrote of his pigeons, "I love them to the extent that I cannot bear to skin and skeletonize them. I have done the black deed and murdered the angelic little Fan-tail Pointer at ten days old."

In the 1870s, the war on animal research heated up in England, and advocates on both sides of the issue sought the support of the country's most renowned scientist. Darwin, however, gave mixed messages. He once referred to physiology as "one of the greatest of sciences." Yet Darwin once complained to a friend that surgery on animals should never be performed "for mere damnable and detestable curiosity."

Ultimately, however, Darwin sided with his fellow scientists. His views on the value of animal research were reflected by a subtle change he made in the second edition of *The Descent of Man*. In the first edition, he wrote, "Everyone has heard of the dog suffering under vivisection, who licked the hand of the operator; this man, unless he had a heart of stone, must have felt remorse to the last hour of his life." Three years later, however, he amended the sentence, adding: "unless the operation was fully justified by the increase in our knowledge." In 1881, he laid his cards on the table, writing in a letter to the London *Times*, "I feel the deepest conviction that he who retards the progress of physiology commits a crime against mankind."

Although Darwin put his weight behind animal research, it was his theory of evolution that muddied the moral waters by undermining the views of seventeenth-century French philosopher René Descartes. Descartes believed that animals were biological robots and their behaviors mere reflexes. Thus scientists could slash and burn as they wished. This perspective was exemplified by the nineteenth-century French physiologist Claude Bernard who wrote, "The physiologist is not an ordinary man: He is a scientist, possessed and absorbed by the scientific idea he pursues. He does not hear the cries of animals, he does not see their flowing blood, he sees nothing but his idea, and is aware of nothing but an organism that conceals from him the problem he is trying to resolve."

Darwin, however, pointed out that because humans and other animals are similar in anatomy and physiology, we also share mental states with them. Modern research in animal behavior has proven Darwin right. In recent years the field of animal cognition has exploded. The list of psychological traits that humans and other species have in common is growing. Scientists have reported that elephants grieve their dead; monkeys and dogs perceive injustice; and cockatiels dance to the music of the Backstreet Boys. There is even a serious journal, *Animal Sentience*, devoted to studies of animal minds. The ethical consequences of Darwin's notion that mental capacities of humans and animals differ by degree rather than kind are inescapable. If animals have perceptions, memories, emotions, and intentions, if they can feel pain and suffer, if they dance and mourn their dead, how can we justify using chimpanzees or cats or even mice in experiments? Is it simply a matter of might makes right?

Animal researchers, thus, face a conundrum. Often, the more similar a species is to humans, the more useful it is as a model to study the biological underpinnings of our afflictions. Because chimpanzees share 98% of their genes with humans, they offer a better model for some human disorders than mice. But because chimps are so similar to us, their use in research is especially problematic. Indeed, the ethical issues are so problematic that in 2013, the National Institutes of Health announced that it was phasing out support for biomedical research on chimps. Hence, the paradox—the more useful a species is scientifically, the less justified is its use morally. This is Darwin's legacy.

Animal activists sometimes claim that modern scientists are no different than their eighteenth-century counterparts in believing animals do not feel pain. For example, in his book *Dominion: The Power of Man, the Suffering of Animals, and the Call to Mercy*, Matthew Scully, wrote, "It remains the working assumption of many if not most animal researchers that their subjects do not experience conscious pain or, for that matter, conscious anything else." Scully is wrong. For an article I was writing on perceptions of animal consciousness, I asked fourteen animal researchers if they thought mice were capable of experiencing pain and suffering. Granted, this is a small sample, but all of them said yes when it came to pain, and twelve of them felt that mice could suffer. In a more systematic survey, all but two of 155 animal researchers believed animals experienced pain.

Most animal researchers today may not view animals as biological robots, but they do not get off the ethical hook as easily as their nineteenth-century predecessors. My friend Phil is a physiologist who studies how cells use glucose and fatty acids—the fuels they need to do their jobs.

Phil is a basic researcher, but he hopes his studies might someday lead to treatments for metabolic diseases such as diabetes. I asked Phil if he ever felt guilty about using mice for his experiments. Only once, he said.

During his post-doc, he was a member of a research team that used knockout mice to discover how cells use energy. Knockout animals are genetically engineered so that some of their genes are turned off. Phil's group used a knockout line of mice to show that a complex molecule called a "transporter protein" enabled fatty acids and glucose to cross into muscle cells where they were used as fuel. Because the transporter gene was inactivated in the knockout mice, the researchers predicted the animals would tire more quickly than normal mice.

Phil's job was to find out how long it took a mouse to run out of gas. A classic way to measure fatigue in rodents is to see how long they can swim. The problem is that air gets trapped in a mouse's fur so they can float around forever, like a kid lying on an air mattress in a swimming pool. "You have to make them swim for their lives," Phil told me. The solution is to rig up a little harness and add just enough weight so the mouse has to swim to keep his head above water.

Phil learned the procedure from a technician who worked in another lab. First, you take a four-inch-diameter graduated cylinder and fill it with water to within a couple of inches of the lip. Then you strap your mouse into the weighted harness, lower him into the water, and start the timer. After swimming a few minutes, the mouse will begin to tire. He will start to sink but then he will fight his way to the surface for a gulp of air. The trick is to let the test continue until you know the animal is going down for good. Then you quickly grab the beaker and dump out the water before the mouse drowns. The guy who taught Phil admitted that a couple of his animals did not make it.

Phil only tested one mouse.

"At some point," he told me, "I could tell that the mouse knew the score, that he had said to himself, 'OK. I know I am going to die, and I just can't do it anymore.' I was supposed to let the test continue to the point where the mouse gives up and sinks and does not try to fight anymore.

But I dumped the water out and the mouse just lay there panting. He was so exhausted."

Phil had had enough. He went to the senior scientist he was working with and told him that he would not take part in the study. The swim tests were reassigned to one of the new graduate students.

Like most scientists who use mice as models of fundamental biological processes, Phil neither likes nor dislikes them. They just happened to be good animals to learn how muscle cells operate. Phil killed a lot of mice over a couple of years with no remorse. Some by cervical dislocation—he would press their heads down with the blunt side of a pair of heavy scissors and break their necks by yanking their bodies backward. Others by decapitation; there was a mouse guillotine in his lab. It looked like a miniature paper cutter.

But when push came to shove, Phil was not a Cartesian. He looked a drowning mouse in the eye and saw a creature with a will to live. He said, "The part that bothered me was that the mouse had given up. I would have loved to be able to finish the experiment, to measure their muscle fatigue. But I could not do it. I didn't want to test their will."

THE MORAL STATUS OF SPACE ALIENS AND HANDICAPPED INFANTS

While scientists do not deny that mice are sentient beings, most animal researchers probably don't spend much time fretting over the morality of their work. But every now and then, something turns your head around. In my case, it was a space alien.

It happened one rainy afternoon when my twin daughters Betsy and Katie, then in kindergarten, were bored and starting to get whiny. To placate them, I rented the movie E.T.: The Extraterrestrial, the Spielberg film about a space alien who becomes stranded in a California suburb. I figured it was just the ticket to keep them occupied for a couple of hours. I needed to finish writing up one of my studies on personality development in baby snakes. The girls were immediately hooked on the flick, and so was I. I stopped working on my research report and watched the movie with them, not knowing it would change my perspective on the use of animals in science.

You probably know the plot. For most of the film, E.T., who has huge puppy eyes and a heart that glows, runs around Southern California with his new human pal, a boy named Elliott. The film ends when E.T.'s mom shows up to fetch her errant son. In the final scene, Elliott reaches out to E.T., pleading, "Stay?" E.T. wistfully shakes his monstrous head, looks deeply into Elliott's eyes, and croaks, "Come?" But, alas, they both know it is not to be. As E.T. creeps up the ramp into the flying saucer for the long ride back to planet Zork, Elliott blinks back a tear, and so did I.

I could not get the movie out of my head, and that evening over dinner, I conjured up a perverse new ending that I tried out on Betsy and Katie. What if, I asked them, the movie ended differently? E.T. asks Elliott to come back to the home planet with him, and just like in the film, Elliott says no. This time, E.T. does not take no for answer. Instead, he growls, grabs Elliott by the arm, and drags the boy kicking and screaming into the mothership. The doors close, and as the movie ends, you hear Elliott shouting "Mommy. Mommy. Help me!" as the ship zooms off into space.

The reason for Elliott's abduction, I explain to the girls, is that a fatal disease is ravishing the population of Zork. Their scientists have come up with a potential cure, and humans, while not as intelligent as the Zorkians, are so biologically similar that they are good animals for testing possible treatments. E.T. was actually in California to collect subjects for these potentially life-saving studies.

"Betsy," I asked. "What do you think? Should E.T. use Elliott in painful experiments that could help save millions of Zorkians?"

"No, Daddy, no!"

"But think about it. Zorkians are a lot smarter than humans. After all, E.T. made a space telephone out of junk, and he has special powers that we humans don't have. He could even make a dead plant bloom."

Katie chimed in, "I don't care, Daddy. It would be wrong for E.T. to put Elliott in a cage and use him for some stupid experiments."

I was not so sure. Like my daughters, I was repulsed by the specter of Elliott sitting forlornly in the alien animal colony where he is poked and prodded and injected with an experimental drug that might save the super-smart Zorkians. But as an animal researcher, I had a problem my daughters did not share. The movie made me realize that the justification for animal experimentation, including my own studies, ultimately rests on the premise that organisms with bigger brains have the right to conduct their research on creatures with less developed mental capacities. Ergo, it is morally permissible for E.T. to haul Elliott off to Zork.

Philosophers have a different version of the *E.T.* dilemma that raises a similar issue. It is called the "argument from marginal cases." Our use of animals in research is predicated on the assumption that nonhuman species lack certain abilities that humans possess—complex emotions, perhaps, or abstract thinking, or the ability to learn language. But what about humans who do not possess these traits? Thousands of children are born each year with severe intellectual impairments making them incapable of ever saying a sentence or thinking about the moral status of mice. The truth is that some people are not as smart as the average chimpanzee, and some humans don't have the mental capacities of a mouse. I cannot see any way to set the ethical bar so it is high enough to exclude all nonhuman animals, low enough to include all human beings, and, yet, be based on morally relevant traits. The ability to feel pain counts; your number of legs does not.

This quandary begs the question: Would it be better to test a drug on an anencephalic infant born without a cerebral cortex—a human infant who is blind, deaf, and incapable of experiencing pain—than on a perfectly healthy mouse? My guts tell me that we should not conduct biomedical studies on profoundly impaired humans rather than mice. But when I posed this question to the philosopher Rob Bass in an email, he wrote back:

My guts deliver a different verdict. It seems obvious to me that research on never-to-be conscious anencephalic children *is* preferable to making mice suffer.

I have found that many of my students also disagree with me: They want to save the mice and conduct our biomedical experiments on death-row prisoners. That's the problem with moral intuition.

WHAT CAN WE LEARN FROM MOUSE RESEARCH?

While a few philosophers might argue that scientists should conduct biomedical research on severely handicapped children, most people would prefer we use animals. But supporters and opponents of animal research bitterly disagree on how much we can learn from mouse research. Geneticists tell you that mouse research has led to breakthroughs in organ transplantation, immunology, our understanding of cancer and cardiovascular disorders, and the causes of birth defects. They want you to know that thirty-five Nobel Prizes in physiology and medicine have been awarded for studies conducted on mice.

On the other hand, groups like the National Anti-Vivisection Society and the Physicians Committee for Responsible Medicine claim that studies on mice are worthless because they are hopelessly flawed and are even detrimental to human health. The truth is probably somewhere in between.

Like it or not, modern biomedical research is built on the backs of mice—many millions of them. As lab animals, mice have a lot going for them. They are fertile, docile, and have fast generation times (one mouse year equals thirty human years). Females become sexually mature when they are only a couple of months old and go into estrus (heat) every four or five days. They produce litters of six to eight pups after three weeks of pregnancy and will happily copulate again just two days after they give birth. Further, because they readily develop spontaneous mutations and their genes can be easily cut up, spliced, and manipulated, they have been compared to the biological equivalent of a Swiss Army knife.

There is another reason that mice make good research animals—most people do not care a twit about mouse rights. In her book *Caring: A Feminine Approach to Ethics and Moral Education*, the philosopher Nel Noddings argues that ethics are based on interpersonal relationships. This explains why she feels no moral obligation to rodents. She writes, "I have not established, nor am I likely ever to establish, a relationship with a rat. . . . I am not prepared to care for it. I feel no relation to it. I would not torture it, and I hesitate to use poisons on it for that reason, but I would shoot it cleanly if the opportunity arose." Most people feel the same way about mice. According to a Zogby poll, 75% of Americans would gladly kill a mouse that showed up in their house. Only 10% indicated they would try to catch the mouse and release it outside, and no one said they would happily let the mouse co-exist in their home.

The transformation of the mouse from pest to pet to model organism began in 1902 when William Castle, a Harvard biologist, obtained inbred mice from a retired Boston schoolteacher to study genetics. Castle was not the first scientist to use mice as subjects. The Austrian monk Gregor Mendel bred mice for his first tentative foray into genetics, only shifting to garden peas after his bishop deemed it unseemly for a man of God to share his living quarters with fornicating animals. The laboratory mouse was officially born in 1909 when a student of Castle's named Clarence Little developed the first purebred line of lab mice. Named DBA (dilute brown non-agouti) for their coat color, DBA mice are still used in biomedical research.

Mouse research mecca is the Jackson Laboratory (JAX), a $400 million nonprofit biomedical research operation. It is based in Bar Harbor, Maine, but has campuses in Connecticut and California. The lab was founded in 1929 by the geneticist and eugenics advocate Clarence Little with financial support from Edsel Ford, the son of Henry Ford. Today, JAX is a rodent factory that produces millions of inbred, mutant, and genetically modified mice each year. Scientists have their pick of 11,000 strains of JAX mice, and, if none of them suit your needs, Jackson scientists will genetically engineer a new strain to your specifications. While some JAX mice are shipped out as live animals, increasingly researchers order their mice as flash-frozen embryos that can be thawed out as needed. The names of the colors of JAX mice remind me of the muted tones on the paint chip samples at Home Depot—"misty gray," "light chinchilla," and "gunmetal."

The variety of JAX mouse infirmities is even more impressive than the colors of their fur. Hundreds of strains are afflicted with rare cancers, others are prone to facial deformities, and some are born with malfunctioning immune systems. There are JAX mouse models for defects of vision, hearing, taste, and balance. JAX mice come with high blood pressure, low blood pressure, sleep apnea, Parkinson's, Alzheimer's, and Lou Gehrig's disease. Researchers trying to cure infertility have their pick of dozens of strains of JAX mice with defective reproductive organs. Then there are the mice that just don't fit in—the obsessive-compulsive, the chronically depressed, the addiction-prone, hyperactive, and schizophrenic mice.

Not surprisingly, JAX mice have been involved in the development of COVID-19 vaccines. Mice are not normally susceptible to the SARS-CoV-2 virus because of differences between men and mice in a receptor in cell membranes called ACE2. However, a sample of frozen sperm of mice developed in 2007 by researchers at the University of Iowa to study SARS was lying around in a JAX freezer. JAX scientists got to work. They thawed out the sperm and were soon able to genetically engineer a strain of mice with humanized ACE2 receptors. Within a couple of weeks they were sending them to researchers to screen potential COVID-19 vaccines and treatments.

Animal research advocates, of course, emphasize the successes. According to the Foundation for Biomedical Research, without animal research, we would not have immunizations for polio, mumps, measles, rubella, or hepatitis. Nor would there be antibiotics, anesthetics, blood transfusions, radiation therapy, open-heart surgery, organ transplants, insulin, cataract surgery, and medications for epilepsy, ulcers, schizophrenia, depression, bipolar disorder, hypertension. They point out that our pets would also suffer—no vaccines for rabies, distemper, parvo, or feline leukemia. Nor treatments for heartworm, brucellosis, cancer, or canine arthritis.

While some scientists have turned to human cells, creatures like zebrafish, and even cybernetic "organs-on-a-chip," the lab mouse remains a staple of biomedical research. Indeed, there has been an upsurge in the demand for mice, and by 2025 the production of lab mice is projected to be a $1.9 billion industry. The growth in the mouse industry is attributable to several factors. These include the need for better mouse models of cancer and autoimmune diseases and the development of techniques like CRISPR which enable scientists to precisely edit strands of DNA.

Mouse researchers claim that almost everything we know about the operation of mammalian genes, including human genes, is rooted in mouse studies. True, the evolutionary paths that led to mice and to men diverged sixty million years ago, and my brain weighs 1,500 times more than the brain of the little fellow that lives behind the filing cabinet in my office. But while we have a different number of chromosomes (he has forty; I have forty-six), we have roughly the same number of genes—25,000, more or less. More importantly, 99.5% of mouse genes have a known human counterpart.

According to Rick Woychik, former president of the Jackson Laboratory, this makes mice the ideal organism that will allow scientists to develop treatments for killers such as juvenile diabetes, breast cancer, and Alzheimer's disease. "It is," said Woychik, "a bench-to-bedside continuum. You start with basic concepts, and then these concepts mature and get translated into clinical concepts and ultimately get delivered as innovative new therapies at the bedside."

JAX researchers are particularly enthusiastic about the new field of personalized medicine. Genes play a role in susceptibility to nearly every disorder from tooth decay to AIDS. Genes also affect how your body responds to medications. Some people receive no benefit from a drug but suffer serious side effects, for example, those four-hour Viagra-induced penile erections that require an immediate trip to the emergency room. Other people, however, experience no side effects and have excellent treatment results from the same medication. The goal of personalized medicine is to tell who will and who will not benefit from a drug. As JAX promotional materials put it, "With ever-increasing precision, we identify the genetic and molecular bases of disease and marshal our strengths in genomics and disease modeling to discover individualized treatment and cures so that medicine is more precise, predictable and personal."

The ethicist Carl Cohen also believed animal research is the key to the advancement of medicine. Cohen was the author of an article in the *New England Journal of Medicine* that is still regarded as the classic defense of animal testing. Cohen wrote:

> Every advance in medicine—every new drug, new operation, new therapy of any kind must sooner or later be tried on a living being for the first time. . . . The subject of that experiment, if it is not an animal, will be a human being. Prohibiting the use of live animals in biomedical research, therefore, or sharply restricting it, must result either in the blockage of much valuable research or in the replacement of animal subjects with human subjects. There are consequences—unacceptable to most reasonable persons—of not using animals in research.

Opponents of animal research frame the debate differently. They throw thalidomide and Vioxx in your face as examples of the failures of tests on rodents to screen drugs that later turned out to be harmful to humans. (Mouse researchers dispute these claims.) They say scientists have exaggerated the contributions of animal research to improvements in our health. The anti-vivisectionists argue that 90% of the decline in the mortality rates for childhood killers such as scarlet fever and diphtheria came before the advent of vaccinations for these diseases. Animal research opponents also argue that improvements to human well-being are really attributable to better nutrition and sanitation. They think studies on mice often lead down blind alleys and impede medical progress.

I support animal research and would like to dismiss the anti-vivisectionists as naïve and uninformed. They do, however, make some legitimate points. For example, even the Pharmaceutical Research and Manufacturers of America admit that only one of every 250 compounds tested on animals will make it to the human trial phase. Then there is the problem of replication of research results. One reason researchers use inbred strains of mice is that they allow scientists in different labs to check each other's findings by independently confirming their results. The comfortable world of mouse researchers was shaken up by an article that appeared in the journal *Science*. Researchers in Portland, Edmonton, and Albany ran eight strains of mice through a series of behavioral tests using precisely the same procedures. The animals in each lab were obtained from the same sources, they were born on the same day, fed the same food, reared on the same light-dark cycle, and put through identical procedures at exactly the same age. The experimenters even wore the same brand of surgical gloves when they handled the mice.

Despite the extreme lengths the researchers took to ensure that the animals were treated the same way, in some tests the mice behaved remarkably differently. A dose of cocaine completely wired the animals in the Portland lab. Their coked-up brethren in Albany and Edmonton, however, showed little response to the drug. The authors concluded that subtle differences between laboratories mean that researchers can arrive at different conclusions even when studying genetically identical animals. I filed the article in my filing cabinet under "inconvenient truth."

There is also the contentious issue of how much we can generalize from mice to humans. Biologically, there are big differences between us and them. We live forty times longer than mice and weigh two thousand times as much. A mouse's metabolism is seven times faster than a person's, and our two species have not shared a common ancestor since the age of dinosaurs.

Writing in the journal *Immunity*, the microbiologist Mark Davis expressed his frustration that studies of inbred mice have failed to produce treatments for people who are already sick. According to Davis, dozens of experimental treatments work on mice with immune system diseases, but few of these results ultimately translate to human therapies. He came to believe that rodents make lousy models for immune disorders.

Ditto neuroscience. Amyotrophic lateral sclerosis (ALS) is a degenerative nerve disease for which there is no cure. Among the dead are the theoretical physicist Stephen Hawking, Yankee slugger Lou Gehrig, and Bob Waters, the football coach at the university where I worked, who, toward the end, was calling plays from a wheelchair, breathing through a respirator. Disheartened that there are no effective treatments he could offer his ALS patients, Michael Benatar, a clinical neurologist from Emory University, read all the published mouse studies of ALS. He was surprised by the results. First, he concluded that most of the research was flawed. Often the samples were too small or the experiments poorly designed. Second, he found nearly a dozen drugs that increased the life spans of mice with the rodent version ALS had no effect when tested on humans. Indeed, one drug that worked in four mouse studies made people with ALS sicker. Benatar compared using mice to study ALS to searching for your missing keys at night under a streetlamp because that's where the light is. The same is true of cancer. Only one in ten cancer treatments that show promise based on mouse studies are ever approved for clinical use with humans.

The list of reasons experiments on mice fail to translate into treatments for human disorders is long—the use of young, healthy, and genetically identical animals of a single sex, insufficient numbers of subjects, researcher expectations, the tendency to only publish experiments that work. The anti-animal research faction, however, should not take too much comfort in the fact that serious scientists are questioning the usefulness of the mouse as a model for human neurological disorders. Some neurobiologists have forsaken mice and have turned to animals whose brains are more like ours—monkeys.

HOW LABELS AFFECT OUR ATTITUDES TOWARD ANIMALS: GOOD MICE, BAD MICE, PET MICE

A recurring theme in anthrozoology is that the ways humans think about animals are mired in an uncomfortable mix of logic and emotion. Some of our decisions about the use of animals in science are perfectly reasonable. Attitudes about animal research depend, in part, on the potential payoff of the experiments, the degree of suffering the animals will experience, and the species used in the research. For example, a 2018 survey in the UK found that 65% of adults approved of animal research for medical purposes if there was no alternative, but only 7% supported safety testing of cosmetics on animals. And 44% of the people agreed with studies using mice while only 14% approved of research conducted on dogs or cats.

At other times, our views about the moral status of animals are morally incoherent. Consider the effect labels and categories have on how we think about mice. I once spent a year as a visiting scholar at the University of Tennessee Reptile Ethology Laboratory. The lab was located in the Walters Life Sciences Building, which was home to hyperactive little marmosets, cooing White Carneau pigeons, beady-eyed albino rats, spiky green tobacco worms, and 15,000 mice. The mice were housed in spotless cedar-smelling rooms in the building's basement where they were cared for by a competent and fully certified staff. But while all the mice in the building belonged to the same species, they were not afforded the same level of moral consideration.

The vast majority of these animals were *good mice*. They were the subjects in the hundreds of biomedical and behavioral experiments conducted each year by the faculty, post-doctoral fellows, and graduate students who worked in the building. Most of these projects were directly or indirectly related to the search for treatments for the various afflictions that affect humans. Though they did not have any say in the matter, these animals lived and died for our benefit. Because the university received grants from the National Institutes of Health, these mice were treated according to the Public Health Service *Guide for the*

Care and Use of Laboratory Animals. Each research project involving the good mice was approved by the university animal care and use committee which was charged with weighing the costs and benefits of the experiments.

There was, however, another category of mice that inhabited the building, the *bad mice.* The bad mice were pests—free-ranging creatures you would occasionally glimpse scurrying down the long fluorescent-lit corridors. These animals were potential threats in an environment where a premium was placed on cleanliness and in which great care was taken to prevent cross-contamination between rooms. These little outlaws had to be eliminated.

The staff of the animal facility had tried several methods of eradicating the bad mice. Snap traps proved ineffective, and the staff was reluctant to use poison for fear of contaminating the research animals. Finally, they settled on sticky traps as the preferred method to capture bad mice. Sticky traps are rodent flypaper. Each trap consisted of a sheet of cardboard about a foot square covered with a tenacious adhesive and embedded with a chemical mouse attractant, hence their alternative name, glue boards. In the evening, animal care technicians would place glue boards in areas where pest mice traveled and check them in the morning. When a mouse stepped on a sticky trap, it would get stuck, and as it struggled, the animal's fur would become increasingly mired in the glue. Though the traps did not contain toxins, about half of the animals were dead when they were found the next day. Mice that were still alive were immediately gassed. Each trap was used only once; mice were not peeled from the trap. The animals caught in the sticky traps suffered a horrible death. I doubt that any animal care committee would approve an experiment in which a researcher requested permission to glue mice to a piece of cardboard. Thus a procedure that was clearly unacceptable for a mouse labeled "subject" was permitted for a mouse labeled "pest."

This paradox was magnified when I discovered where the pest mice came from. The building, it seemed, did not have a problem with wild rodents. But in a facility housing thousands of small creatures, leakage is inevitable. Thus virtually all the bad mice were good mice that had escaped. The animal colony manager told me, "Once an animal hits the floor, it is a pest." *Poof!* Its moral status evaporates.

The moral status of a mouse at the University of Tennessee depended on whether it was labeled a subject or a pest. I was quick to criticize this seemingly arbitrary distinction until I realized that the same theme was playing out in my own home. For our son's seventh birthday, I kidnapped a mouse who was destined to become a meal for IM, the two-headed black rat snake, and I gave him to Adam as a birthday present. Adam named his mouse Willie and we set up a home for his new pet in a cage in his bedroom. We liked Willie. He was quiet and affectionate. But mice have short life spans, and one morning Adam woke up and found Willie lying dead on the bottom of his cage. We held a family discussion, and the children decided a funeral ceremony would be appropriate. We put Willie in a little box and buried him in the flower garden with a piece of slate for a headstone. We stood around saying nice things about him, and Betsy and Katie cried a little; it was their first encounter with death.

A couple of days later Mary Jean, a neatnik, discovered mouse droppings on the kitchen floor. She looked at me and said, "Kill it." That night I put a dab of peanut butter in a snap trap which I placed on the floor between the refrigerator and the stove. I found the mouse the next morning. It was a clean kill. This time, there was no funeral. I tossed the little guy's body into the bushes not far from Willie's grave, and it struck me that the labels we assign to the animals in our lives—research subject, pest, or pet—affect how we treat them more than the size of their brain or whether they experience happiness.

ARE MICE ANIMALS?

In 1876, the British Parliament enacted the world's first law regulating the use of animals in research. The United States caught up ninety years later.

The events that precipitated congressional action in the United States were a pair of articles on dogs. The first was a 1965 *Sports Illustrated* story about Pepper, a Dalmatian who disappeared from her yard one afternoon, apparently abducted by a dealer who provided animals to laboratories. Pepper's distraught owners finally located their dog, but only after she had been euthanized during an experiment in a New York hospital. A year later, an article appeared in *Life* magazine titled "Concentration Camp for Dogs." Again, the story focused on the horrid treatment of family pets who wound up as laboratory subjects. Members of the House and Senate were bombarded with letters from constituents worried that their cats and dogs might suffer a similar fate. For a couple of months, Congress received more mail about animal research than the two great moral issues of the time, Vietnam and civil rights. The House and Senate quickly enacted the Animal Welfare Act of 1966. (It was not until 1974 that the government took steps to protect the rights of human research subjects.)

The bureaucratic gyrations of the Animal Welfare Act exemplify the convoluted ways humans think about other species. Perhaps the strangest aspect of the legislation concerns a straightforward question—what is an animal? The act's definition of the term *animal* starts reasonably enough: "Animal means any live or dead dog, cat, nonhuman primate, guinea pig, hamster, rabbit or other such warm-blooded animal, which is being used, or is intended for use for research, teaching, testing, experimentation, or exhibition purposes, or as a pet." The smoking gun is in the next sentence. "This term excludes: birds, rats of the genus *Rattus* and mice of the genus *Mus* bred for use in research. . . ."

That's right, according to Congress, mice are not animals. Neither are rats or birds. This means that over 99% of the animals used in research in the United States are not covered under the main federal animal protection legislation. (Mice and other vertebrates used in research at institutions that receive grants from the National Institutes of Health are covered under a separate set of guidelines.) Federal Judge Charles Richey called the mouse/rat/bird exclusion in the Animal Welfare Act arbitrary and capricious. He was right. For instance, the legal definition of the word *animal* means that a researcher who unobtrusively videotapes the sexual behavior of white-footed mice (genus *Peromyscus*) has to jump through all the federal legal hoops. His friend down the hall who delivers electric shocks to brain-damaged lab mice (genus *Mus*), however, is completely exempt from the regulations.

It is instructive to compare how the Animal Welfare Act treats mice, a species most people do not like, with our best friend, the dog. Because mice are not animals, they have no standing under the law. End of story. Dogs, in contrast, are singled out for special treatment. They are entitled to a daily dose of "positive physical contact with humans" (I think this means play). Ironically, because the act applies to dead as well as living animals, dead dogs have more legal protection than live mice. (A footnote in the Animal Welfare Act, however, exempts dead dogs from the minimum cage-size requirements.)

Because mice, rats, and birds are not considered animals under the act, we don't know how many animals are used in research each year in the United States. I can tell you that in 2018, exactly 59,401 dogs, 18,691 cats, 171,406 guinea pigs, and 70,797 monkeys were used in biomedical and behavioral experiments. But these numbers don't include the most commonly used species of lab animals. If you toss in the excluded species, most authorities place the total number of animals used in research somewhere between seventy and twenty million. These numbers, however, are almost certainly a gross underestimate. In a 2021 paper published in *Scientific Reports*, Dr. Larry Carbone used the Freedom of Information Act to find the number of rats and mice used each year in large American research facilities. Based on this data, he calculates that 111 million rodents were used in biomedical research between 2017 and 2018. And he estimates that about 99% of them were mice.

The Animal Welfare Act has been tweaked over the years. The most important amendments were added in 1985 when Congress took on the issue of which studies are worth doing and which ones are not. In Great Britain, every animal experiment must be approved by the Home Office in London. Congress took a different route and placed the responsibility for ensuring the ethical treatment of lab animals on the institutions where the research was conducted. It directed each institution to establish a local Institutional Animal Care and Use Committee, or IACUC. These committees must have at least three members, though most have more, including a veterinarian and a member from outside the research facility who represents the interests of the community.

Serving on an IACUC is a tough job. Animal care committee members at major universities can spend hours each week poring over the fine print in proposals that can run fifteen or twenty pages. Every couple of months, they get together and play God. The members thrash out which proposals to approve, reject, or request more information about. The lives of animals hinge on their decisions as do scientific careers. Being an IACUC member is a good way to lose friends. But can these committees accurately weigh the benefits of an experiment against the costs in terms of animal suffering?

JUDGING THE JUDGES: HOW GOOD ARE THE DECISIONS OF ANIMAL CARE COMMITTEES?

Some years ago, I received a phone call from Scott Plous, a social psychologist from Wesleyan University who studies the psychology of decision making. Both of us were interested in how people think about other species, and we had once run into each other while handing out surveys to activists at an animal rights demonstration in front of the Capitol in Washington, D.C.

"Hal, have you ever considered doing a study where you would ask different animal care committees to evaluate the same proposals?" he asked.

"Of course," I said. After all, it would be nice to know that the system Congress set up to ensure the welfare of research animals worked—that the animal care committees at the University of Texas and Johns Hopkins University would make the same decisions about the same experiment.

"But Scott, it would be impossible. Scientists are busy. You would never get them to cooperate."

Scott disagreed. He thought committees would participate if you offered them money they could use to enhance animal care at their university. I was skeptical but I said, OK, count me in. Scott pitched the idea to the National Science Foundation, and to my astonishment, they approved our proposal. Scott was right—by offering the school extra funds for animal care, we recruited fifty randomly chosen university IACUCs to participate in our study. Indeed, the committees were enthusiastic about the project. In the end, roughly 500 scientists, veterinarians, and community members took part in the study—nearly a 90% response rate.

Each committee chairperson sent us three research protocols that had already been reviewed by their committee. After removing identifying information, we sent them off to be re-reviewed by the committee at another university. The proposals ranged from studies of how bats locate water holes to the development of eating disorders in mice. In all, the 150 proposed projects involved over 50,000 animals, mostly mice and rats, but also a smattering of other species—chimpanzees, frogs, buffalo, egrets, pigeons, dolphins, monkeys, sea turtles, bears, lizards, you name it. When the data were in, I flew up to Connecticut to help Scott crunch the numbers and figure out what they meant. I had served as an animal care committee member, and I was sure that there would be reasonably high levels of agreement between the first and second IACUCs.

I was so wrong.

There are moments of truth in science. For me, it is the millisecond gap between the time you push the enter key on your computer and the results flash on the screen. Scott and I were sitting in his office, our eyes on the screen. I was antsy, feeling a little rush of anticipatory adrenaline, like an offensive lineman waiting to hear the quarterback yell "Hutt!"

Scott pushed the button. The numbers popped up. Our jaws dropped.

About 80% of the time, the second committee made different decisions than the first one. Our statistical analysis indicated that the committees might as well have made their decisions by flipping a coin.

The research oversight system was clearly inadequate. Why, I wondered, should it be OK to suspend rats in ice water in California but not in New York? In retrospect, I should not have been surprised to find that the decisions of animal care committees are wildly inconsistent. It is harder than you think to tell good from bad research. In his novel *Zen and the Art of Motorcycle Maintenance*, Robert Pirsig laid the issue out nicely: "But, if you can't say what Quality is, how do you know what it is, or how do you know that it even exists?" For a scientist, this is a question that can keep you up at night.

Our finding that different animal care committees often make different decisions was not an anomaly. Studies showing inconsistencies in peer review judgments of quality in science go back forty years. They include studies of ratings of grant proposals, journal article submissions, and even the quality of research abstracts submitted for presentation at anthrozoology conferences. It seems that scientists are not particularly good at discerning the quality and importance of research.

Put simply, the system Congress enacted to oversee the treatment of research animals is fraught with inconsistencies. Why are white-footed mice but not lab mice covered by the Animal Welfare Act? Why are dogs but not cats entitled to a play session every day? Why can a project be given full approval by one animal care committee and flat out rejected by a different one? Unfortunately, these nagging problems give credence to the charges by anti-vivisectionists that when it comes to animal research, the fox is guarding the henhouse.

What can we do about this situation? For starters, Congress should extend the Animal Welfare Act to include all vertebrate species— mammals, birds, reptiles, amphibians, and fish. (British animal research regulations even extend to octopuses.) Our research suggests that most scientists also want mice, rats, and birds covered under the Animal Welfare Act. Three-fourths of the nearly 300 animal researchers who participated in our study said they disagreed with the Animal Welfare Act's definition of the word *animal*.

Of course, we could just dump the present system. We could either let scientists conduct animal research without any external oversight or we could throw a pair of dice to decide which animal experiments should be conducted. Both alternatives are unacceptable. Some animal rights activists argue for a third alternative. They would have us ban animal research altogether. But people who oppose all animal experimentations are up against their own inconsistencies and paradoxes.

USING ANIMAL EXPERIMENTS TO SHOW THAT YOU SHOULD NOT CONDUCT ANIMAL EXPERIMENTS

The argument against animal research is based on the premise that mice and chimpanzees fall within the sphere of moral concern but that tomato plants and robotic dogs do not. That's because animals have mental traits that plants and machines don't possess. For example, the influential philosopher Tom Regan restricted the possession of rights to species that possess consciousness, emotions, beliefs, desires, perceptions, memories, intentions, and a sense of the future. But how do we know which animals have these attributes? The answer, of course, is animal research.

The legal scholar Steven Wise, founder of the Nonhuman Rights Project, is one of the few animal rights advocates who has seriously grappled with the moral implications of differences in mental capacities among the species. In his book *Drawing the Line: Science and the Case for Animal Rights,* Wise developed a 0 to 1.00 "Practical Autonomy Scale" on which species are rated according to their cognitive abilities. The rankings are based on Wise's review of scientific studies of animal behavior and cognition. Humans are assigned a 1.00 on the scale; chimpanzees .98; gorillas .95; African elephants .75; dogs .68; and honeybees .59. Wise argues that creatures scoring above .90 (great apes and dolphins) are clearly entitled to basic legal rights while animals with scores below .50 are not. The strength of this approach to animal ethics is that an animal's moral standing is based on evidence rather than naïve conjectures about their abilities or how much we like them. For instance, after reviewing the science, Wise concluded that Alex, the language-trained African gray parrot, earned an autonomy score of .78 which is higher than a dog.

There is, however, a paradox associated with Wise's empirical approach to animal rights—you need to conduct animal research to determine if it is immoral to use a species in animal research. Wise, for instance, assigns dolphins an autonomy scale score of .90, which puts them in the highest category of nonhuman creatures that deserve legal rights. He writes, "Dolphins have concepts and spontaneously understand pointing, gazing, and the holding up of replicas. They instantly imitate actions and vocalizations." His assessment of the cognitive abilities of dolphins is largely based on the findings of a University of Hawaii psychologist named Lou Herman. Over three decades of research on captive animals, Herman demonstrated that dolphins have extraordinary memories, can read human gestures better than chimpanzees, and have such sophisticated linguistic skills that they will correct your grammar.

Given that Wise's case for dolphin rights is based on Herman's research, you might think Wise would be a fan of these studies. Wrong. Indeed, he vehemently argues that Herman's dolphin research was unethical, that Herman exploited his research animals, and that the animals were prisoners. The irony, of course, is that without these studies of dolphin cognition, Wise would not have been able to argue that the mental abilities of dolphins are comparable to chimps and that dolphins, therefore, are entitled to legal rights.

What about mice? Where do they fall on the autonomy scale? Wise does not mention them in his book, so I sent him an email:

Professor Wise:

Where do mice rank on your scale? They are, after all, the most common mammal used in research.

Wise replied that the omission of mice was simply a matter of time constraints. The autonomy rankings, he said, are based on an objective assessment of the available evidence for the mental capacities of each species. This task requires tracking down the latest research reports and interviewing leading scientists who have studied the behavior and cognitive abilities of each species. Wise said that in the cases of apes and dolphins, the data fit his preconceptions. On the other hand, honeybees scored much higher than he ever anticipated. The evaluation process takes roughly three months for each animal. There are only twenty-four hours in a day and thousands of species.

DO MICE EXPERIENCE EMPATHY?: THE MCGILL PAIN STUDIES

Steven Wise admits we don't know enough about the mental abilities of most species to accurately place them on his moral status scale. This would seem to mean we need more, rather than less, animal research. Some of these studies would certainly discover that nonhuman animals possess unexpected capacities. Researchers at McGill University's Pain Genetics Laboratory, for example, conducted a series of experiments they claimed showed mice are capable of empathy. While I am not convinced that *mucine* empathy is analogous to the human experience of empathy, their findings raise a thorny ethical issue.

The purpose of the study was to discover whether mice would react to suffering experienced by other mice. The researchers used several procedures to induce pain in the animals. Most of the mice were subjected to the unfortunately named "writhing test" in which they were injected in the stomach with a diluted solution of acetic acid. In others, their hind paws were injected with an irritating liquid. Animals in the third group were subjected to a "paw withdrawal test" which involved measuring how quickly a mouse would lift its feet from a hot surface. If I calculated correctly, the research involved over 800 mice.

Did the mice feel each other's pain? The short answer is yes. Animals injected with acetic acid writhed more when tested near another writhing mouse than when tested alone. But here is the interesting part—pain contagion only occurred when the other mouse was a relative or a cage-mate. The mice showed no signs of empathy in the presence of strangers that were in pain.

The next question was: How does a mouse know if his cage-mate is suffering? Do they see an agonized look in their pal's eyes or hear their ultrahigh-frequency moans? Or perhaps a mouse in pain emits a fear pheromone. The researchers checked each of these possibilities by systematically disrupting the sensory systems of the mice. Vision was easy. The scientists just put an opaque screen between two writhing mice. Eliminating their sense of smell, however, was grizzly. After injecting a mouse with a local anesthetic, they would flood each nostril with a caustic chemical that fried the smell receptor cells in the animal's nose. This procedure permanently destroyed its ability to ever smell anything again. To eliminate hearing, they injected mice with a chemical called kanamycin every day for fourteen days. Two weeks later, the mice were deaf.

From a scientific perspective, the experiment was a success. The researchers discovered that the mouse empathy communication system relies on visual cues. Mice deprived of their senses of smell and hearing remained empathetic. The mice which were blocked from seeing their suffering compatriots were not.

But was the study ethical? Pretend, for a moment, that you were a member of the McGill University Animal Care and Use Committee charged with approving or rejecting the mouse empathy study. How would you have voted? Did the results of the experiments justify the pain and suffering of the animals?

Make your decision: approve or reject.

For me, this is a tough one. The research was well done, and while most scientific articles are never read by anyone, the results of this study were published in the journal *Science* and garnered worldwide publicity. Further, the researchers made a reasonable argument that the pain was relatively mild and short-lasting.

But I vote to reject.

The reason is that I find listening to Chris Stapleton plaintively wailing "Tennessee Whisky" one of life's pleasures, and I love the toasty aroma of French bread in the oven. Hence, I do not like the idea of deafening and erasing the sense of smell of hundreds of mice. (I might approve the study if the researchers would agree to dump the sensory deprivation experiments.)

When I read the research report, my first thought was, "These guys are in deep shit." I figured they would be getting death threats from the lunatic wing of the anti-research movement. I was wrong. The radical Animal Liberation Front (ALF), a group that promoted the harassment of animal researchers, prominently featured the McGill mouse pain study on its website as evidence that men and mice are kindred spirits. Even some scientists who normally oppose experiments involving the infliction of suffering in research animals seemed to tacitly approve of the study. Marc Bekoff, for example, is an eminent ethologist and a powerful voice for animal protection. He argues that scientists should not conduct research on animals that they would not do on their own dogs. Thus I was surprised to find that he used the mouse pain study in his book *Wild Justice: The Moral Lives of Animals* as evidence that even rodents experience sophisticated emotional states.

Jonathan Balcombe is also an animal protectionist with an animal research background. (His doctoral dissertation was on the behavior of bats.) The author of several books on the inner lives of other species, Jonathan opposes all invasive and painful research on other species and is a popular speaker in animal protection circles. Articulate, thoughtful, and calm, Jonathan is the perfect face for a movement that is too often stereotyped as a band of wild-eyed fanatics. Given his opposition to invasive animal research, I was puzzled when Jonathan used the McGill pain study results to argue that mice have emotions at a conference presentation. So I called him up to find out how he would have voted if he were on the McGill animal care committee.

"Of course, I would have voted to reject it," he said.

"But don't you find it paradoxical that so much of what we know about the mental abilities of animals is based on research that would not be permitted if you had your way and experiments on captive animals were abolished?" I said.

Jonathan was ready for that question. It turns out that he gets asked this a lot during the Q-and-A sessions following the talks he gives on university campuses. Some smart-assed grad student in neuroscience will stand up and say something like, "Dr. Balcombe, you say you oppose animal research, yet much of your argument rests on experiments that have harmed animals. Isn't that a contradiction?"

This is not an ethical dilemma for Jonathan. "I hate these studies," he tells the audience. "If I had my way, we would not allow some of the experiments that I use to show that animals have feelings. But the fact is that they have already been done, and they do shed light on the question of animal consciousness. So I am going to keep using them. "

It is obvious to me that Jonathan has thought a lot about this issue, but I am surprised when he brings up the Nazi medical experiments. The reason is that, prodded by the mouse pain study, I have also been thinking about them. Dr. Sigmund Rasher, a German physician, immersed prisoners in Dachau in frigid water for extended periods to see how long pilots could survive if they were downed in the icy waters of the North Sea. Nearly a hundred people died during the study. By some accounts, this research remains some of the best information we have on the effects of hypothermia on the human body. Some medical ethicists believe that because the data derived from the horrors of the Dachau and Auschwitz experiments have already been collected, we honor the dead by using this information to save human lives today—even if it was obtained unethically. Others argue that the data are morally tainted, ill-gotten gains that should not be used under any circumstance. Similarly, some animal activists believe the results of experiments on animals are also ill-gotten gains. They believe, for example, it is immoral to take medicines that have been tested on animals.

Are the results of the McGill pain experiments or, for that matter, studies of language learning in captive chimpanzees, also ill-gotten gains that should not be used, even to make the case against animal research? Jonathan is not losing much sleep over this one. When it comes to the campaign against animal research, he admits that he has reluctantly become a utilitarian. "I am willing to use any available evidence to plead the case of the animals. Whatever works," he tells me. But then he adds, "Within reason."

DO NUMBERS COUNT? MICE IN THE AGE OF COVID-19

But reason can be elusive in debates over animal research. As is often the case with our attitudes toward other species, public opinion regarding the use of animals as research subjects is completely inconsistent. I think the argument for the use of nonhuman animals in biomedical research is much stronger than, say, for eating meat or recreational hunting. Yet only 51% of Americans support animal research while 80% of my fellow countrymen support the right to hunt and kill animals, and 95% support the right to consume animal flesh.

As I write this, 4.5 million people have died from COVID-19. Scientists are in a high-stakes arms race with the virus as new variants crop up—some of them increasingly contagious and deadly. Is there a morally coherent approach that would allow us to weigh the interest of humans and nonhuman species when making decisions about animal research? When I asked Justin Goodman if all bets are off during a global epidemic, he instantly said no. Justin is vice president for advocacy and public policy of the anti-animal research organization the White Coat Waste Project. He tells me animal research—even with mice, even in the search for improved COVID-19 vaccines—is morally wrong, scientifically unjustified, and a waste of money. But he admits that he is the exception—that most of the 49% of Americans who tell pollsters they oppose animal research will change their minds in a flash as COVID-19 hits close to home.

For a philosopher's perspective, I turned to Princeton University's

Peter Singer—arguably the world's most influential ethicist. Singer is the author of *Animal Liberation*, the book that inspired the modern animal rights movement. His moral philosophy is rooted in utilitarianism—the idea that we should aim to do the greatest good for the greatest number. And in his moral calculus, animals count. Singer is generally opposed to animal research, at least experiments that we would not be willing to conduct on humans.

I was unclear if his thinking on the animal research issue had changed in the face of COVID-19, so I asked him in an email. Within a couple of hours, I had his response:

Hal, what the pandemic changes regarding the ethics of animal experiments, is not the utilitarian model but the numbers that get fed into it. The use of animals in research is justifiable when the benefits are clearly outweighed by the costs—where the interests of all sentient beings are given equal consideration, and the benefits are discounted by the odds against the research project achieving those benefits.

Then he got down to the numbers . . .

A project with, let's say, a 1% chance of benefiting seven billion humans is more likely to outweigh the certain suffering of a given number of animals than a project with a 10% chance of benefiting only seven million sentient beings, assuming that the costs to the animals and the benefits to the humans are similar in both cases.

But my animal rights friend Janet does not buy Singer's argument a bit. When it comes to animal research, she is an absolutist. I knew she was opposed to animal studies, but I wanted to know if her ideas had changed at all in the face of COVID-19, so I gave her a call. Our phone conversation went like this:

HAL: Janet, COVID-19 has already killed three million people. Jackson Labs has produced lines of genetically engineered mice that can be used to develop vaccines and treatments. Even if only 5% of mouse experiments help produce a human cure, millions of human lives would be spared. Has your moral equation changed in the face of the present pandemic?

JANET: No. It does not change for me. I don't believe humans have any greater value than the animals that would be tested and killed in these experiments. We get these viruses from other species because of the ways we relate to the nonhuman creatures around us—because of the way we treat animals. I have grieved and grieved and grieved about the animals that have died to create these vaccines. I literally start crying when I think about it. But I have also grieved and grieved for the people who have died.

HAL: Janet, let's pretend for a minute that you were on an animal care committee and had to approve or disapprove of a study in which a couple of hundred genetically engineered mice will be used to develop an improved COVID-19 vaccine. Would you vote no?

JANET: I would never serve on one of those committees.

HAL: I know. But is it safe to say that you would oppose the study even in the face of the very real COVID-19 pandemic?

JANET: Yes! Always!

Then we chat a while. We say we look forward to when things are back to normal, and we can get together again. And she casually mentions she is relieved that her mom will be getting her COVID-19 vaccine shot next week.

The Cats in Our Houses, the Cows on Our Plates

ARE WE ALL HYPOCRITES?

> If some animals count for something, which animals count, how much do they count, and how can this be determined?
>
> —ROBERT NOZICK, PHILOSOPHER

> Stop smirking. One of the most universal pieces of advice from across cultures and eras is that we are all hypocrites, and in our condemnation of others' hypocrisy we only compound our own.
>
> —JONATHAN HAIDT, PSYCHOLOGIST

If you visit Seattle, don't miss the Pike Place Market. Every year, ten million visitors flock to the flower stalls, bakeries, fresh produce stands, and the assorted cheese, candy, mushroom, fruit, and gourmet salami shops. The biggest draw is the Pike Place Fish Market where men wearing rubber boots and gray hoodies confidently fling fifteen-pound king salmon twenty feet through the air into the cradled arms of another man in a hoodie standing by the cash register. The crowd loves to see the big fish fly. They laugh and take pictures. I have seen it myself. I laughed and took pictures too.

The American Veterinary Medical Association decided that a fish-catching demonstration would be a terrific team-building exercise for the

10,000 veterinarians and paraprofessionals who would be attending the organization's 2009 conference in Seattle. PETA was not amused. In an article in the *Los Angeles Times*, a PETA campaign manager named Ashley Byrne was quoted as saying, "Killing animals so you can toss their bodies around for amusement is just twisted. And it sends a terrible message to the public when vets call it fun to toss around the corpses of animals." The media played it for laughs, and my first thought was Ashley needs to get a life. But then PETA issued a statement saying that the crowd in the Pike Place Fish Market would not be laughing so loud if the guys in gray hoodies were throwing around the bodies of dead kittens. That's when I realized PETA was right. Why should people think it is funny to play catch with a dead fish but not the carcass of a cat?

WHY OUR ATTITUDES ABOUT THE TREATMENT OF ANIMALS ARE SO MORALLY INCOHERENT

Elizabeth Anderson, author of *The Powerful Bond Between People and Pets: Our Boundless Connections to Companion Animals*, is troubled by this kind of moral inconsistency. She is, for example, puzzled by pet owners who wear fur coats. Anderson wrote, "How a person who has ever loved or kissed a puppy or a kitten can turn a blind eye to the anal electrocution of a mink or the head-bashing of a seal pup, I doubt I will ever understand." She should not be surprised that a person who melts at the sight of a kitten can also love the luster of mink. Glaring inconsistencies occur even among many people who take the rights of animals seriously. The social psychologist Scott Plous found that 70% of animal activists who felt that the use of animals for clothing should be the top priority of the animal rights movement admitted that they wore leather products.

Psychologists have long known that our words and deeds are often at odds. A widely accepted theory of attitudes is called the "A-B-C model." It holds that attitudes have three components: Affect (how you feel emotionally about an issue); Behavior (how the attitude affects your actions); and Cognition (what you know about the issue). Sometimes the components work together.

Take, for example, the philosopher Rob Bass. Rob's life was going along just fine until he came across an article by an ethicist named Mylan Engel twenty years ago. Engel made a logical argument against eating animals that Rob—to his surprise—found compelling. Rob figured there had to be a flaw in Engel's logic and he spent the next three weeks trying to disprove the argument. After a month, he gave in. Once he was convinced that Engel was right (a cognitive change), he knew he had to quit eating meat (a behavioral change). A few weeks later he walked by the college cafeteria and caught a whiff of burgers frying on a grill. His response was immediate and visceral—"Yuck. That smell is *so* disgusting!" (an affective change). Reading Engel's article had started Rob on a cycle in which his behavior, thinking, and emotions reinforced each other. Now Rob is a strict vegan. He and his wife, Gayle, are opposed to the exploitation of animals of all kinds, and Rob teaches animal rights in his college ethics classes.

Rob, however, is the exception. Most people seem blithely untroubled by the contradictions in their attitudes about animals. The *Los Angeles Times* once commissioned a survey in which a random sample of American adults were asked whether they agreed or disagreed with the statement "Animals are just like people in all important ways." The paper reported that 47% of respondents agreed with the statement. I was skeptical about the results, so I decided to see how my students responded to the item. I gave a survey to one hundred of them that included the *Los Angeles Times* item as well as a dozen other questions about the treatment of animals. My skepticism was unjustified. Exactly 47% of the students also agreed that animals were just like people in all important ways. But their belief that humans and animals are equal had little effect on their attitudes about other uses of animals. Half of the students who said that animals were "just like people" favored using animals in biomedical research, 40% thought it was OK to replace diseased human body parts with organs taken from animals, and 90%

regularly dined on the creatures they believed were "just like humans in all important ways."

How can people maintain such blatantly contradictory opinions? Most people's views about the treatment of other species exemplify what psychologists call "non-attitudes" or "vacuous attitudes." These are superficial collections of largely unrelated and isolated opinions, not the coherent belief system that we see in people like Rob who have thought deeply about moral problems involving animals. The ethical issues associated with our relationships with other species are complex and most people—even those who say they are animal lovers—are somewhere in the middle. For instance, when asked in a National Opinion Research Center survey how they felt about animal testing, only one in five adults said they had strong opinions one way or another about the topic, and 15% had no opinion at all.

While there are plenty of exceptions, the evidence indicates that, compared to their beliefs about other social issues, the majority of people do not get in much of a twit over the treatment of animals. Since 2001, the Gallup Organization has regularly surveyed American adults about their most important social and political concerns. The list is long and always includes topics such as abortion, gun control, climate change, poverty, health care, and crime. Concern for the treatment of animals never makes the list. The Humane Society of the United States commissioned a survey in which people were asked which national animal protection group does the most to protect animals; half of the respondents could not name a single organization that promotes the interests of animals. Finally, a survey of those who boycotted consumer products found that only 2% did so out of concern for the treatment of animals. The truth is that, aside from personal pets, the treatment of animals is not a high priority for most people.

If you really want to know how people feel about the treatment of animals, follow the money. According to Andrew Rowan, president of WellBeing International, Americans donate about $5 billion a year to animal protection and control organizations each year. However, that's less than 1% of total annual philanthropy in the United States. Further, it pales when compared to the money we spend to kill animals—$170 billion on meat; $27 billion on hunting supplies, equipment, and travel; and $9 billion to kill animal pests. We spend, of course, vastly more— about $100 billion—on our pets than we contribute to organizations that promote the welfare of animals we do not personally know.

This is perfectly consistent with several fundamental principles of human nature. One is the well-established evolutionary psychology axiom that family comes first—and pets are now considered family members in most American homes. Another is a phenomenon the University of Oregon cognitive psychologist Paul Slovic calls "psychic numbing"—the fact that the larger the tragedy, the less people care. Mother Teresa nicely captured Slovic's idea when she said, "If I look at the mass I will never act. If I look at the one, I will." Slovic's research supports these sentiments. In one study he found that individuals would donate twice as much to save one sick child as they would to save a group of eight sick children. In another, he reported that donations to the Swedish Red Cross to aid refugees from the Syrian civil war temporarily jumped a hundredfold following the widespread publication of a photograph of the body of a three-year-old Syrian-Kurdish boy washed up on the Mediterranean shore. As Joseph Stalin is reputed to have quipped, "One death is a tragedy, a million deaths is a statistic."

Slovic refers to human indifference in the face of overwhelming numbers as "the collapse of compassion." But when it comes to the treatment of animals, not everyone suffers from compassion collapse. Ten million Americans are members of the Humane Society of the United States. The ASPCA claims two million members, and PETA nearly seven million. Many of these people don't just contribute money—they take action. One of the first projects I undertook in anthrozoology was a series of interviews with animal rights activists. I focused on people working at the grassroots level, the foot soldiers—not movement leaders, philosophers, or celebrities. I wanted to know what kinds of people are drawn to the animal rights movement, why they became involved in animal protection, and how this moral commitment had affected their lives.

It turned out that three out of four animal rights activists were women, and most of them were politically liberal, well-educated, solidly middle class, and primarily white. Nearly all of them, of course, had pets. Animal activists had come to the movement via different paths, but the most common thread was moral shock. For Katherine, a nurse, the shock was caused by a single photograph.

"What drew you to the animal liberation movement?" I asked.

"A picture on a PETA poster. I can still remember the picture of that little monkey. They had severed his nerves, and he couldn't use his arm. They had taped the other arm and made him use the handicapped arm."

"You still remember what that picture looked like?"

"Oh yes," she said. "This monkey had really beautiful eyes and it looked like it had been crying. It makes me feel like crying." At this point, Katherine began to cry softly, and she said, "I didn't realize I was so emotional about it until I started talking about it."

Opponents of animal rights will run into someone like Katherine and assume that all animal activists are hypersentimental types who prefer the company of animals to people. This is a mistake. Many of the activists I have spoken with have a firm rational basis for their opposition to the exploitation of animals. One woman who was very conversant with the nuances of the intellectual case for animal rights said she resented it when people called her "soft-hearted." She said, "To pass off all the years I have been thinking through these issues as being 'soft-hearted' is really condescending."

ANIMAL LIBERATION AS RELIGION

As a group, animal rights activists are not very religious, at least not in a conventional sense. Shelley Galvin and I found that only 30% of participants at a large national animal rights protest indicated that they were members of traditional religious denominations, and about half of them said they were atheists or agnostics. But as with other moral crusades, the animal liberation movement has religious elements. Animal activism can give your life meaning and purpose. When I asked Phyllis how important the animal rights movement was to her, she blinked and seemed puzzled, like the answer was self-evident. "It is my life."

Mark was a retired policeman. He told me he had been clinically depressed before he and his wife became involved in animal protection. He felt that the animal rights movement saved him. He said, "It's one of those things that happen once in a lifetime that make you happy doing what you are doing. It affects your whole existence. We are just totally happy."

You get the sense when you talk to Mark that, like Saint Paul on the road to Damascus, the scales were suddenly pulled from his eyes and he saw the light. Brian, a self-confessed agnostic, said to me, "Sometimes I laugh at myself, and feel like I know how a 'born again' probably feels. Just like me, their beliefs affect every aspect of their lives." Another activist said, "I have grown to respect Jesus in a very different way. I think if Jesus were alive today, he certainly would be a vegetarian. I think he would be an animal rights activist."

Animal rights activists and religious fundamentalists are alike in another way—they see moral issues in terms of black and white rather than shades of gray. Shelley Galvin and I gave animal activists a psychological scale designed to assess individual differences in ethical ideologies. Seventy-five percent of animal activists fell into the "moral absolutist" category compared to only 25% of a group of college students. Individuals with this ethical stance believe moral principles are universal and that doing the right thing results in happy endings.

THE PSYCHOLOGICAL CONSEQUENCES OF TAKING ANIMALS SERIOUSLY

Big things happen when you decide to take animals seriously. First, you have to change your life. All the animal activists I have met have taken steps to bring their behavior in sync with their beliefs. Some take baby steps, others giant steps, and some are more successful than others. Marie was the biggest failure; she only lasted two weeks. During a lunch break at her first (and last) animal rights conference, Marie had a Big Mac attack and snuck over to a McDonald's for a burger. That was the end of animal rights for her. She was, however, the exception. Of activists I surveyed who were attending a large demonstration in Washington, D.C., 97% had changed their diet (though a quarter of them still ate some meat), 94% purchased consumer products labeled "cruelty-free," 93% boycotted companies that tested products on animals, 79% said they avoided clothes made from animals, and 75% had written letters about the treatment of animals to newspapers or legislators. Their beliefs and behavior reinforced each other. As Gina told me, "The more I got involved, the more my diet changed. And the more my diet changed, the more involved I got."

The moral commitment of activists shows up in many ways. Some refuse to kill animals not normally regarded as pets. One man told me he had spotted a copperhead in his garden. A year before, he said he would have grabbed a hoe and killed the snake—but now he carefully nudged it back into the woods. Bernadette was an IBM executive who lived a conventional upper-middle-class existence, complete with a husband, two kids, a minivan, and a dog. What made her different from the other women in her subdivision was that she would, literally, not kill a flea.

"Bernadette," I asked. "Can you give me an example of how your views on animal rights affect you on a day-to-day basis?"

"Well," she said, "I don't use toxic chemicals on my dog to get rid of fleas. Instead, I try to pick them off and put them outside. I know they do not feel pain or anything, but I feel it is important to be consistent. If I draw the line somewhere between fish and mollusks or something, it isn't going to make sense."

But then the roaches showed up. "We recently annihilated the roaches in our house," she said. "But before we resorted to Terminix, I walked around for a week trying to telegraphically tell the roaches, 'You have invaded my territory and we are going to take drastic action.' In my fantasy, I was hoping they would magically disappear." Unfortunately, they didn't.

Bernadette was up against "the activists' paradox"—*the greater your moral clarity, the harder it is to be morally consistent.* Small things can become an issue. For Gina, even eating plants posed a dilemma. She sometimes wondered if a fruit and nut diet was ethically preferable to eating plants like carrots that do not survive harvesting. Roy's passion was church-league softball. After months of searching, he found an OK (but not great) synthetic glove. But he could not find a decent ball that was not made of leather. Fortunately for Roy, a lot more products are now available for people seeking a cruelty-free lifestyle. These include cruelty-free synthetic softballs and vegan-friendly condoms.

In his book *The Happiness Hypothesis: Finding Modern Truth in Ancient Wisdom,* Jonathan Haidt argues that the keys to a happy life are a sense of virtue and moral purpose, a feeling of enlightenment, volunteerism, and solidarity with a group with shared core values. Many animal activists have these, so you would think they would be among the happiest people on earth. This was certainly true of the woman I met at a protest against the mistreatment of captive black bears in roadside zoos. "I just feel sorry for people who don't have something like this in their lives," she said. It was also true of Mark, the policeman, who said about his commitment to animal rights, "It's one of those things that happens in one's lifetime that makes you happy doing what you are doing. It does affect your whole existence. I'm just totally happy."

But other activists paid a heavy price for their moral clarity. For instance, allegiance to animal liberation can alienate friends, family, and lovers. While most Americans *say* they support the notion of animal rights, in reality people are often uncomfortable around individuals who take the issues very seriously. As an activist named Alan told me, "My friendships have suffered a great deal. Nobody understands what I am doing, and I feel a lot of defensiveness from them. I have completely lost my closest friend of ten years, and a lot of it had to do with animal rights."

Commitment to animal protection can even affect marriages. For Hugh and Lydia, the cause of animals was a joint commitment, a common focus that made their marriage stronger. They cooked vegan meals together, went to the same conferences, and gave each other feedback on articles they were writing about the treatment of animals. But it doesn't always work this way. Animal activism destroyed Nancy's marriage. Her husband of ten years was a military man who was hostile to her increasing dedication to animal liberation; he wanted her to play the role of a good army wife.

"So eventually," she told me, "I had to make a choice." She chose animals.

Fran and her husband were on a similar collision course.

"How does your commitment to animals affect your relationships with other people?" I asked her.

She sighed. "My husband and I have lots of fights about it. He is a meat-eater and thinks that people who wear fur are not any worse than people who eat meat—that is not true. Over the years we have been married it has gotten worse. For the past two years, he has thrown my mail away because I send so much money to animal organizations."

I put the odds that they are still together at zero.

Lifestyle conflicts fell particularly hard on single activists who were looking for like-minded dating partners. With a gender ratio of three females to every male activist, this burden falls particularly hard on women. Sandra told me she had given up hope of finding a long-term partner. "Just going out to dinner becomes an ordeal," she said sadly.

Other problems come with being a moral crusader. Sometimes the burden just gets too heavy. I asked Lucy, a special education teacher, if people think she's crazy because of the way she lives her life. "No," she said. "I don't think most people feel that I'm nuts. But sometimes I think I'm nuts. Because I drive myself crazy about it. It dominates my life. Sometimes I think I can't take it anymore. So I say to myself, I'm going to back off a bit; I'm going to loosen the rope a little. I'm going to let myself not be Jesus for a minute and be a normal human being."

When I posed the same question to Judy, a vegan from northern Georgia, she spoke longingly of her pre-activist life. "Sometimes I will be driving down the street, and I will see someone that looks happy, and they don't have any burning issue. And I just think, God, I would love to be a normal person. But now that I am involved in animal rights, I can't envision myself leading a normal life like other people, like my family."

Adding to their psychological burden is the fact that animal activists are constantly bombarded with reminders of animal cruelty—the meat counter at the grocery store, the smell of grilled flesh when they walk past a Burger King, the woman wearing a fur coat at the airport, the barrage of fund-raising messages from animal protection organizations that flood into their email inbox pleading:

> Your donation will help us stop the baby seal hunt! Let's put an
> end to puppy mills! Shut down factory farms!

Sometimes moral commitment can become overwhelming. Susan suffered from chronic insomnia from dreams that were haunted by images of animal mistreatment. Maureen and her husband were forced to declare bankruptcy because they had donated all their money to animal rights organizations. And Hans, a sixty-two-year-old German-born businessman, was suffering from compassion fatigue. "I have come to near emotional collapse," he told me. "I am burning out. My life is so full of animal rights now I have no time for anything else anymore. I have thrived on this in the past. But this year it came to the point when I said, 'I can't do it anymore. I just don't have the strength.'"

Like most individuals who take moral issues to heart, animal rights activists march to the beat of a different drummer. But the vast majority of activists are not fanatics. Most of the activists I have met over the years have been intelligent, articulate, friendly, and completely sane. Nonetheless, it can be hard to have a meaningful conversation with true believers, impossible to find a middle ground. Good luck explaining to Lucy why you think some animal research might be justified. I asked her if she ever had moments of doubt—if she ever thought that maybe there are circumstances in which it might be OK to use a heart valve taken from a pig heart to save a child.

"No," she said. "I definitely have the sense that what I am doing is right. And if you argue with me I am not going to listen. Because I *know* I am right."

That's a conservation stopper.

ANIMAL RIGHTS AND TERRORISM

Occasionally, moral fanaticism hits close to home. One of my favorite places to hike in western North Carolina is the Pisgah National Forest not far from my house. On the afternoon of Sunday, October 25, 2020, a man named Tyler Mayo was walking with his dog Bobby-Joe and a friend on the Foster Creek Trail when he suddenly felt an intense stabbing pain in his left foot. He looked down and discovered his foot was impaled on a device called a "nail trap"—a piece of plywood in which someone had hammered twelve two-inch nails. They had carefully placed the trap on the trail, points facing up and covered with leaves. Mayo told the *Asheville Citizen-Times,* "We were having this amazing time, the dogs were having fun, and then I just stepped on this nail trap. And it drove two nails super deep into my heel." He added that he had a hard time pulling the nails out of his foot before heading to the hospital.

The trail is located in a remote section of the national forest which is used almost exclusively by hunters. Clearly, someone was trying to make a statement. Bear season had just opened, and the perpetrator had crudely scrawled PETA on the plywood. The next day, however, PETA vehemently denied any involvement in the incident. I believe them. This kind of attack is not their style, and the nail board could just as easily have injured an animal. Indeed, PETA offered a $1,000 reward for information leading to the arrest of the perpetrator.

Often, people identify the animal liberation movement with terrorists in ski masks who break into laboratories in the middle of the night to free the mice. This is a mistake. The Center for Strategic and International Studies is a Washington, D.C.–based think tank that monitors domestic terrorism in the United States. According to their analysis, the face of terrorism has changed over the past twenty years. Left-wing attacks, most of which involved the destruction of property related to animal research and farming, peaked between 2000 and 2005. Over the next fifteen years, the percentage of left-wing attacks, including those related to the use of animals, dropped from about 73% in 2002 to 6% in 2019. (During the same time, right-wing domestic terrorist activities increased from 30% of attacks to over 90%.)

David Grimm covers animal rights and welfare issues for *Science* magazine. He tells me that direct attacks on researchers and labs have dwindled in recent years—that animal rights groups have shifted their tactics from the barricades to the halls of Congress. For example, the new White Coat Waste Project is a coalition of anti-vivisection liberals and fiscal conservatives. The organization has ended a number of government-funded animal research projects by painting them as a misuse of taxpayer dollars.

But while attacks on researchers have declined, they still raise an important issue. Is violence ever justified in the name of a moral crusade? And, if so, under what circumstances? Take the case of the neuroscientist David Jentsch who headed a research lab at UCLA. In the middle of the night on March 7, 2009, Jentsch awoke to the blare of a car alarm. He looked out his bedroom window and saw that his Volvo was in flames. He ran outside and grabbed a garden hose. Jentsch lives in one of those Los Angeles neighborhoods that are prone to runaway wildfires. The branches of the tree above his car were already burning when the fire department showed up. The whole neighborhood could have been taken out if the firemen had been caught in a traffic jam.

Two days later, a group called the Animal Liberation Brigade released a communiqué that said, "David, here's a message just for you. We will come for you when you least expect it and do a lot more damage than to your property. Wherever you go and whatever you do, we'll be watching you as long as you continue to do your disgusting experiments on monkeys."

Jentsch was not completely surprised. Nearly a dozen UCLA researchers had been subjected to animal rights terrorist attacks. (The animal liberation underground describes these attacks as "direct action.") While most victims of these incidents decided to lay low, Jentsch fought back. He formed an organization called UCLA Pro-Test to defend animal research on campus, and the group staged a rally in support of animal experimentation. This did not make him friends among the animal rights fringe. He began getting emails with messages like this:

> David Jentsch, I want all your children to die of cancer, and I want you to watch them die. I hope you die a horrible death too.

Jentsch was undeterred. In 2011, he was given the Award for Scientific Freedom and Responsibility by the prestigious American Association for the Advancement of Science. Today, as a member of the psychology department at Binghamton University, Jentsch uses mice to uncover the genetic and neurobiological underpinnings of drug and alcohol addiction.

According to social scientists, terrorists are usually moral absolutists motivated by a combination of idealism, anger, religious zeal, and the natural human penchant to place the blame for injustice on villains. Gerard Saucier of the University of Oregon and his colleagues analyzed the thinking of a dozen kinds of militant extremists. Common elements run through these disparate groups—the belief that peaceful tactics don't work, the ends justify the means, utopia is around the corner, the need to annihilate evil, demonizing the opposition, and the moral vision as war.

You saw all of these in a tiny but violent wing of the animal liberation movement—the arsonists and bomb throwers, the spray painters, the lab animal "liberators"—the fanatics. You also heard it in the words of individuals like Jerry Vlasak, a physician who is "press officer" of the North American Animal Liberation Front. Vlasak once told an Australian television reporter, "Would I advocate taking five guilty vivisectors' lives to save hundreds of millions of innocent animal lives? Yes, I would."

Antiabortion extremists in the United States have killed at least eleven people. No one has died in an animal rights attack. Yet James Jarboe, former head of the Domestic Terrorism Section of the FBI, once testified before Congress that the animal rights and environmental extremists were among the most serious domestic terrorism threats. Why did biomedical researchers and not hunters or slaughterhouse owners become the main target of the terrorist wing of the animal liberation movement? After all, the numbers of animals used in research are minuscule compared to the ten billion animals jammed into factory farms or the untold millions of wild animals killed or wounded by hunters each year. The overwhelming majority of animals used in research are rats and mice, creatures most people would not hesitate to personally kill on sight (or at least pay someone else to do it for them). And, of all ways that humans use other species, experiments aimed at developing treatments for diseases that affect people and their pets is probably the most justified. Animal research is certainly more defensible than eating creatures because they are tasty or shooting them for recreation.

The biggest predictor of whether researchers were targeted by animal activists was the species they work with. Most of David Jentsch's experiments involve mice and rats. But what drew the arsonists to his house was that he occasionally used vervet monkeys in his studies. Researchers who study chickens, lizards, gypsy moths, tobacco worms, trout, spiders, parrots, mice, and rats were rarely bothered by animal activists. Rather, the targets were usually scientists who worked with either monkeys or pet species (usually cats). In terms of the amount of suffering associated with animal research, this does not make any sense. At UCLA, for example, about 75,000 mice were used in research every year compared to several dozen monkeys. Three-fourths of the attacks posted on the ALF webpage were directed at primate researchers even though monkeys and apes made up less than 0.3% of animals used in research at the time. In contrast, only 9% of attacks described in the communiqués were directed at scientists experimenting on rats and mice—the animals used in over 95% of biomedical studies.

Like most University of California researchers who were attacked by anti-vivisectionists, Jentsch studied brains. His research concerned the neural mechanisms underlying schizophrenia and the effects that drugs like angel dust, ecstasy, cocaine, and nicotine have on nerve cells. Why did animal liberation terrorists focus more attention on scientists seeking treatments for mental illness, drug addiction, and blindness rather than researchers working on diseases like cancer or AIDS? One reason is these studies often involved primates because their brains are similar to ours. The ultimate goal of violence-prone animal liberationists, however, was the elimination of all animal research. (A fruit fly geneticist at the UC Santa Cruz was targeted.) However, activists in the shadow world made a strategic decision to concentrate on researchers who study a handful of species the average person empathizes with. Pictures of cute monkeys make for better fund-raising brochures than photographs of beady-eyed albino rats.

Given the toll that the threats, the emails, and the car bombing had on his life, David Jentsch had a surprisingly positive attitude toward local animal rights activists. After the attack, Jentsch met with local animal protectionists to try to develop what he calls a talking relationship with them, to try to get a conversation going. He told me most of the activists were reasonable individuals who opposed the harassment of researchers. They did not appreciate the attention the media was giving to the Animal Liberation Front. They felt the voices of reason were being drowned out by the crazies. Jentsch said that in the Los Angeles area, only a tiny group of zealots were responsible for the death threats, the obscene emails, and the fire bombings.

But he added, "It only takes one person to plant a bomb."

MORAL CONSISTENCY AND ANIMAL LIBERATION PHILOSOPHY

Putting an incendiary device on someone's doorstep is animal liberation run amok. But most animal activists are not nut cases. To appreciate the larger issue raised by direct action on behalf of animals, it helps to understand the ethical foundations of the modern animal liberation movement.

Much like journalism, the study of ethics boils down to who, what, and why questions—who is entitled to moral concern, what obligations do we have to them, and why one course of action is better than another. The technical literature concerning our obligations toward other species is vast, complicated, and, for the most part, boring. The philosophical case for giving moral status to animals has been made by Aristotelians, feminists, Darwinians, Christian right-wingers, and post-modernist leftists. However, the major intellectual paths to animal liberation lie in the two classic approaches to ethics—utilitarianism and deontology. Utilitarians believe that the morality of an act depends on its consequences. Deontologists, on the other hand, argue that the rightness or wrongness of an act does not depend on its consequences. They hold that ethics are based on universal principles and obligations (the term *deontology* comes from the Greek word for obligation–*deon*). For example, you should keep your promises not because bad things will happen if you break them, but because you made them.

The application of utilitarian principles to the treatment of animals was originally made by the eighteenth-century philosopher Jeremy Bentham. He argued that acts should be judged on their consequences—specifically, the degree to which that they increase pleasure and decrease pain. His twist was to insist that other species be included in the moral calculus. He wrote, "The question is not, 'Can they reason,' nor 'Can they talk,' but, 'Can they suffer?'" Peter Singer used this line of thinking as the cornerstone in his 1975 book *Animal Liberation: A New Ethics for Our Treatment of Animals*, which jump-started the contemporary animal liberation movement.

Ironically, while his book is often called "the bible of the animal rights movement," Singer's argument for animal liberation is not based on the idea that animals (or humans for that matter) have inherent rights. Rather, his case is based on simple fairness. Singer lays his position out in a single sentence: "The core of this book is the claim that to discriminate against a being solely on account of their species is a form of prejudice, immoral and indefensible in the same way that discrimination on the basis of race is immoral and indefensible." He refers to bias toward the interests of your species and against members of other species as "speciesism," which he regards as morally repugnant as racism and sexism. Singer's argument rests on the notion that all sentient creatures (that is, organisms capable of experiencing pleasure and pain) have the same stake in their own existence and that suffering is the ultimate moral leveler. "From an ethical point of view," he writes, "we all stand on an equal footing—whether we stand on two feet, or four, or none at all."

The deontological argument for animal liberation was laid out by the late Tom Regan in his book *The Case for Animal Rights*. Regan started with the idea that humans and some animals deserve moral consideration because they are the "subjects of a life" and, therefore, have inherent value. By this, he means they possess memories, beliefs, desires, emotions, a sense of the future, and a sense of their own existence over time. Regan believed that if a creature has inherent value, it is wrong to treat it as a mere thing to be used or discarded.

I am often irritated with the balky computer on my office desk. Regan would say that it is morally permissible (and I would add, psychologically rewarding) to heave it out the window which is on the third floor. But, according to Regan, it would not be ethical for me to toss an irritating student out the window because they want to argue with me over a grade. Nor would he approve of throwing Tilly out the window because I grew tired of her nagging me to rub her belly while I am trying to write about the ethical problem of having a predator for a pet. Regan reasoned that, as subjects of a life, both the argumentative student and my cat have fundamental rights that my computer does not possess. And, most importantly, they have them in equal measure. Among these include the right to be treated with respect and the right not to be harmed.

Regan and Singer differ on some issues. For example, they disagree as to why it would be wrong for me to toss a student and/or a cat out a third-floor window. Singer would say it is because they would suffer, not because they have inherent rights. And, while Singer is not—at least in principle—opposed to painlessly taking a human or a nonhuman life in some circumstances, Regan is. The two philosophers, however, agree on most of the big issues. They both acknowledge that there are important differences between humans and other animals, but they believe that these are not relevant to whether a creature deserves moral consideration. The logical extension of both Regan's rights argument and Singer's utilitarian logic is that we should not eat animals, hunt them, or otherwise avoidably cause them to suffer. Practices such as factory farming, keeping animals in zoos, or trapping them for their fur are immoral under both positions.

CAUGHT IN THE GRIP OF A THEORY: ANIMAL ETHICS AND FOOLISH CONSISTENCIES

Inevitably, ethics involves drawing lines. Singer originally drew the line "somewhere between the shrimp and the oyster" while Regan set the bar at the level of mammals and birds at least one year old. (He bent the rules a bit by saying that basic rights also extend to humans less than a year old.) Singer and Regan both recognize that humans live in the real world, not in an imaginary moral ether populated by intellectual airheads. Thus they are willing to make the occasional compromise to accommodate common sense. Both, for example, implicitly recognize that some species warrant more concern than others. Singer has invested more energy into promoting an international campaign to gain legal standing for our closest relatives, apes, than he has to banning mousetraps. And Singer has acknowledged that in rare cases, biomedical research using monkeys might be justified. Regan concluded that if four normal humans and a golden retriever are in a lifeboat that can only carry four, the dog goes overboard. He wrote, "Death for the dog is not comparable to the harm that death could be for any of the humans."

But what happens if you refuse to draw moral lines in the sand? Ralph Waldo Emerson famously wrote, "A foolish consistency is the hobgoblin of little minds." In animal ethics, foolish consistencies are exemplified by Joan Dunayer, author of the book *Speciesism*. By insisting on a combination of animal rights literalism and uncompromising adherence to moral consistency, she constructed a set of impossible ethical standards. She also illustrated the consequences of taking moral logic too far.

Dunayer would, of course, consider me a speciesist because I eat meat. The surprise is her vicious attack on the intellectual elite of the animal liberation movement such as Peter Singer and Steven Wise. She even goes after animal rights groups such as PETA because they do not measure up to her standards of ideological purity. Dunayer, for example, adamantly opposed any effort to reduce animal suffering by substituting a cruel practice with a less painful, but still harmful, alternative. Thus she denounced PETA because of their efforts to pressure the fast-food industry to make life better for chickens on factory farms. Bigger cages to Dunayer are unacceptable; it is empty cages or nothing. She fumed at Tom Regan for arguing that the dog should be booted out of the hypothetical life raft ahead of a person.

Peter Singer merits a special place on Dunayer's list of animal rights cop-outs. She disapproves of Singer's efforts to single out chimpanzees and gorillas as candidates for legal standing. (I assume she would also disapprove of Singer's statement that he would not have much compunction about swatting a cockroach; insects don't suffer much, he says.) Dunayer was not happy when Singer argued that the deaths of the 3,000 humans who died on September 11, 2001, were a much greater tragedy than the deaths of the thirty-eight million chickens killed in American slaughterhouses that day. Dunayer disagrees. In her eyes, chickens deserve more moral consideration than humans. She writes:

> Singer's disrespect for chickens is inconsistent with his espoused philosophy which values benign individuals more than those who,

on balance, cause harm. By that measure, chickens are worthier than most humans, who needlessly cause much suffering and death (for example, by wearing animal-derived products).

Philosophers have a phrase for what happens when people take logic to bizarre extremes. They say you are "caught in the grip of a theory." Dunayer falls into the grip by making two assumptions that seem reasonable until you play them out. The first is that all creatures capable of experiencing pain should be treated equally. The second is that all it takes to experience pain is the simplest of nervous systems.

In Dunayer's own words, here are some of the logical consequences of these seemingly innocuous assumptions:

- "Because all sentient beings are equal, we're perfectly entitled to save the dog over any of the human beings." (p. 97)
- "Wasps need a legal right to life." (p. 141)
- "Our moral obligations need to include insects and all other beings with a nervous system. . . . These animals include comb jellies, cnidarians such as jellyfishes, hydras, sea anemones, and corals." (p. 127)

Joan Dunayer lives in a moral universe which should cause even hard-core animal activists to shudder. Can a reasonable person really believe, as Dunayer apparently does, that one should flip a coin when deciding whether to snatch a puppy or a child from a burning building, or that duck hunters should be sentenced to life in prison?

The problem for animal rights extremists is that Dunayer is right. If you take the charge of speciesism literally, if you refuse to draw any moral lines between types of animals, if you really believe that how we treat creatures should not depend on the size of their brains or the number of their legs, you wind up in a world in which, as Dunayer suggests, termites have the right to eat your house.

HOW SHOULD A GOOD PERSON ACT?

I hate my Inner Lawyer. He usually pops up when I am taking a shower in the morning, or driving on some mountain road with the radio turned off, or when I am kayaking on a stretch of river that does not demand my attention. He's my Jiminy Cricket, my Obi-Wan Kenobi. He asks me inconvenient questions. According to Harvard neuroscientist Joshua Greene, he lives in a little section of my brain behind my eyebrows called the DLPFC—the dorsolateral prefrontal cortex. Using a functional MRI to image the brain, Greene found that our DLPFC lights up like a Christmas tree when we try to think logically through tough moral issues. My Inner Lawyer showed up a couple of days ago when I was hiking in the Smokies.

I.L.:   Hal, it's me—your Inner Lawyer.

HAL: Go away.

I.L.:   Just listen for a second. Pretend that it is 1939 and you are living in the quaint little village of Dachau outside of Munich. Every day you see the smoke from the chimneys behind the fence of the new "camp" and know that Hitler's goons are working the ovens over-time to exterminate Jews and homosexuals. Rumor has it that Nazi doctors are even conducting painful medical experiments on some of the prisoners. Your friend Heinz asks you to help him plant a bomb under the SS guards' barracks. "By killing them we will save many lives," Heinz says. "We will send a message to the world." Do you think you should help Heinz blow up the barracks and perhaps save thousands of people, Hal?

HAL: I wouldn't have the guts.

I.L.: I know that. But just pretend you are both a brave and a good person.

HAL: *sigh* . . . OK, a good person with courage would be justified in taking direct action to help prevent genocide, even if it meant killing a barracks full of Nazis.

I.L.: I agree. Now assume you have read books by the intellectuals of the animal rights movement. They have convinced you that speciesism is the moral equivalent of racism, that the suffering of a monkey in a laboratory is not morally different than the suffering of a human child. You also know that 60,000 monkeys a year are used in research in the United States. And, you agree with the spokesman of the Animal Liberation Front that killing just one or two primate researchers might bring all these studies to a halt.

Now, answer this question, Hal. Would a good and brave person like you be justified in firebombing the home of a scientist who was addicting monkeys to cocaine to study the brain chemistry of addiction?

HAL: No, it would be against the law.

I.L.: But murder was against the law in Nazi Germany, and you said that was OK.

HAL: That was different.

I.L.: Why?

HAL: Because keeping monkeys in a lab and even killing them for science is not the same as killing Jews in a concentration camp.

I.L.:   But if you *believe* there is no morally relevant difference be-
        tween a person and a monkey, wouldn't you be justified in
        harming monkey researchers?

        *Hmmm . . . I am thinking that I.L. has me by the balls. But
        then I am saved by a dim recollection of a lecture I heard back
        in school on Immanuel Kant's theory of ethics.*

HAL: Gotcha. Kant argued that you should always act in such a way
        that you would want everyone to act in the same situation. I
        would not want to live in a world where every whacked-out
        moral crusader with a gun would be allowed to shoot people
        they thought were doing harm to their pet cause—old-growth
        forests, fetuses, or frogs. The world would be chaos.

I.L.:   But Hal, you did say it would be justified to blow up the Nazi
        concentration camp guards. How do you know when it is OK
        to take an illegal action and when it is not?

HAL: You just KNOW. It is common sense!

I.L.:   So, you think that when it comes to killing someone you
        should rely on your own common sense, your personal moral
        intuition? Would Kant go for that?

HAL: Get out of my life, asshole.

IN MATTERS OF MORALS, YOU CAN'T TRUST YOUR HEAD . . . OR YOUR HEART

My Inner Lawyer has raised the question at the center of all human morality, not just animal ethics: How do we know what is right? There are

two places where we can turn to for moral guidance—our head and our heart. The problem is that you can't rely on either of them.

First, head. Cognitive psychologists have repeatedly demonstrated that human thinking is, in the words of the behavioral economist Dan Ariely, "predictably irrational." Researchers have identified dozens of types of bias that unconsciously warp the way we think. They have great names: the Lake Wobegone Effect, Myside Bias, the Gambler's Fallacy, the Barnum Effect, Naïve Realism. The list goes on.

Dunayer's conclusion that a spider and a human child have the same moral status is at the same time logical and ridiculous. It illustrates how pure reason can lead us to completely warped ethical standards, that even when we apply the rules of logic, things can go awry when making ethical decisions. My philosopher friend Rob Bass disagrees. He says that if you correctly apply formal deductive logic to premises that are true, you will always end up with a correct conclusion. In theory, he may be right. However, psychologists have repeatedly found that humans vary greatly in their ability to think rationally about moral issues. Further, there is abundant evidence that there is almost no relationship between the sophistication of a person's ethical thinking and how they actually behave.

Even Tom Regan and Peter Singer, both first-class intellects, get into trouble by taking moral consistency too far. For example, in his lifeboat scenario, Regan concluded that the dog goes overboard first. Then he took it a step further and said that you should toss a million dogs overboard if it would save a single human. But, at the same time, Regan argued that it is wrong to sacrifice a million mice for biomedical research that might ultimately save millions of human children.

Logic also led Peter Singer to some conclusions that most people would find unnerving. In *Practical Ethics,* he shows that the logical upshot of his utilitarianism is that it might be OK to euthanize a seriously disabled infant if the child's mother might subsequently give birth to a healthy child. And Singer once raised the possibility that sexual interactions between humans and animals are not necessarily harmful to man nor beast. While his remarks were taken out of context, his comments were met with howls of protests by both the press and animal advocates.

The adherence to cold logic in moral decision making has led some philosophers to conclude that it is preferable to use impaired human infants rather than monkeys in biomedical experiments, that arson is a legitimate agent of social change, and that the life of an ant and an ape are of equal moral value. So much for relying on our heads when it comes to thinking about animals.

What about our hearts? Is moral intuition better than logic at resolving the moral conundrums in our dealings with other species?

Unfortunately, no. If anything, in matters of morality, our hearts are even more prone to error than our heads. Intuition and its handmaiden, common sense, are subjected to the whims of a host of morally irrelevant factors—how big an animal's eyes are, its size, whether it was the mascot of your high school football team, and the evolutionary history of our species. Moral intuition told my friend Sammy Hensley that his hounds did not mind spending their lives chained to a doghouse and Japanese fishermen that there is nothing wrong with slaughtering dolphins because they are fish. My moral intuition tells me that it is OK to eat meat (particularly if it is labeled "cruelty-free") but my friend Al's moral intuition tells him that meat is murder. For thousands of years, it was common sense that slaves were property and homosexuality was a crime against nature. Moral intuition told the 9/11 airplane hijackers and the arsonists who planted the firebomb under David Jentsch's car that they had the moral high ground.

SEEKING THE MORAL HIGH GROUND

I am confused and need another opinion on the "oughts" of ethics and animals, so I send an email to my friend Gayle Dean, an animal advocate who takes ethical issues seriously.

Gayle, from an animal liberation perspective, is there any difference between the 9/11 terrorists and "direct action" ALF types who attacked scientists? After all, both groups were completely convinced that they have the moral high ground.

She writes back:

There is a big difference between groups that are convinced they have the moral high ground and the ones that actually have the moral high ground. The difference lies in the truth of the matter. During slavery, many people risked their lives to illegally help slaves because they were convinced they had the moral high ground. In fact, they did have the moral high ground. The same for those who helped Jews escape from the Nazis.

It is after midnight. Tilly has fallen asleep on the rocking chair in my office. I am tired, but I write back.

Gayle, I agree with you about the Nazis and slavery. But here is my question—How can we be sure what the moral truth is in these matters? Doesn't it boil down to personal opinion, your own moral intuition?

The next morning she responds:

I agree that it is difficult to know the truth of the matter. But moral truth certainly does NOT boil down to personal opinion!

I'd like to think she is right, but the longer I study human-animal interactions, the more I have my doubts.

Jonathan Haidt says we are all hypocrites. After twenty years of studying how people think about animals, I have come to believe he is right. You do, of course, run into the occasional exceptions like Lisa who is a vegan and does not take antibiotics or let her cat outdoors where it might enjoy stalking birds. But the vast majority of humans are inconsistent in their attitudes and behavior toward other species, often wildly so. What are we to make of this?

In the 1950s, the social psychologist Leon Festinger proposed one of the most influential theories in psychology—when our beliefs, behavior, and attitudes are at odds, we experience a state that Festinger called "cognitive dissonance." Because dissonance is uncomfortable, people should be motivated to reduce these psychic conflicts caused by inconsistency. We might, for example, change our beliefs or our behaviors, or we might distort or deny the evidence.

The environmental ethicist Chris Diehm (who is a vegan) also believes that people are motivated to seek moral consistency. When he points out the moral inconsistencies about the treatment of animals to people, he says they often make an effort to become more consistent, or at the very least, develop some sort of rationalization to resolve the contradictions in their beliefs and behavior. He writes:

> We recognize that our relationships to animals take widely disparate, apparently contradictory paths: We have cats in our houses but cows on our plates. When people have this inconsistency pointed out, they try to make sense of it, or remove it to the point where they are comfortable with it. The drive for consistency seems to be a good thing, and exposing inconsistencies is a deep motivator for moral reflection and development.

Chris is a philosopher. He is impressed by the need for humans to achieve logical coherence in their beliefs and behaviors. I am a psychologist. I am more impressed by our ability to ignore even the most blatant examples of moral inconsistency in how we think and behave toward animals. In my experience, most people—be they cockfighters, animal researchers, or pet owners—remain stubbornly oblivious (other than an occasional uncomfortable laugh) when you point out the paradoxes and inconsistencies in our personal and cultural treatment of animals.

Dealing with the Carnivorous
Yahoo within Ourselves

ARE WE ALL HYPOCRITES?

> The fact that you can only do a little is no excuse for doing nothing.
>
> —JOHN LE CARRÉ, NOVELIST

> Uncertainty is the only certainty there is, and knowing how to live with insecurity is the only security.
>
> —JOHN ALLEN PAULOS, MATHEMATICIAN

The central character of the Nobel Prize-winning author J. M. Coetzee's novel *Elizabeth Costello* is a visiting scholar who delivers a series of lectures on the moral status of animals at a major university. After one of her public talks, an audience member raises her hand:

> Are you not expecting too much of humankind when you ask us to live without species exploitation, without cruelty? Is it not more human to accept your own humanity—even if it means embracing the carnivorous yahoo within ourselves?

Good question. Coping with our inner yahoos is a central theme of ethics, psychology, and religion. The yahoo goes by different names. Freud called it the Id. George Lucas called it Darth Vader. When the Apostle Paul wrote that the spirit is willing but the flesh is weak, he was warning about the yahoo. George Jones sang about it in "Almost Persuaded." Evolutionary psychologists trace its origins to the Stone Age, and neuroscientists say it divides its time between your frontal lobes and your limbic system.

Jonathan Haidt, the psychologist who has explored the moral ramifications of the yahoo more than anyone else, compares it to an emotional elephant being ridden by a rational rider. The elephant is large and usually calls the shots, albeit unconsciously. The rider is weaker but smarter. With practice, it can exert some control over the elephant. I have argued in this book that the paradoxes characterizing our relationships with other species are the unavoidable result of the perennial tug of war between the rational you and the yahoo within. But what are the implications of living in a morally convoluted world in which consistency is elusive if not impossible? Do we throw up our hands in despair?

Not always, says philosopher/psychologist/neurobiologist Joshua Greene. Greene likens our moral processing system to a camera that comes with built in presets like "portrait" and "landscape" as well as a fully manual mode. Manual mode requires more expertise but it makes for better photographs in difficult lighting situations. Similarly, when it comes to moral reasoning, our default mode is unconscious and rapid emotion—Haidt's elephant. But when the situation calls for it, we can sometimes flip into manual—the rational rider.

Greene nicely sums up his ideas in an interview with National Public Radio's behavioral science correspondent Shankar Vedantam: "When it comes to the ethics of everyday life, the basic things that everybody should know not to do—the lying, stealing, cheating kinds of things—we want to have automatic settings that just say, nope, nope, nope, nope, nope, can't do that. But then, sometimes life is complicated. And sometimes there are difficult tradeoffs. And then that's when you want to shift into manual mode."

But in matters of animals and ethics, switching from auto mode to manual mode does not necessarily benefit animals. The social psychologists Lucius Caviola and Valerio Capraro asked people to imagine difficult situations in which they had to save either a human or a chimpanzee with a very high level of intelligence. In half the cases, their subjects were told to rely on logic and reason in making their decision (Greene's manual mode). The other half of the subjects were told that emotion leads to good decisions and they were instructed to let their feelings guide their choice (Greene's auto mode). The researchers anticipated that when primed to think rationally, the subjects would be less likely to prioritize humans over animals than when they were told to listen to their hearts. They were wrong. Indeed, they found the opposite—the more people relied on their heads than their hearts, the more they valued human life over the life of an animal.

Humans (and possibly, only humans) have the capacity to struggle with the carnivorous yahoo within ourselves. During my studies, I have met many people who have made the switch from auto mode to manual mode in their thinking about animals. These people work for animals in different ways and on different scales. Most of them do small things that help animals. Some of them have cut back on their meat consumption or adopted a shelter dog. They may have donated money to PETA or the World Wildlife Fund or pulled over to the side of a highway and moved a box turtle in the middle of the road to safety.

Others worked for animals on a larger scale. Michael Mountain is one of them.

SAVING ANIMALS ON A GRAND SCALE: A GLOBAL KINDNESS REVOLUTION

A man walks into a bar . . .

The bar was in the Sheraton Hotel in Raleigh, North Carolina, where I was attending a conference on human-animal relationships. The man was in his early sixties, tall, wiry, reddish hair, neatly trimmed beard, outdoorsy—a ruddy Abraham Lincoln. He looked around and saw that the only empty seat in the place was next to me.

"Do you mind if I sit here?" English accent. Oxbridge.

"No. Have a seat. I'm Hal Herzog."

"Michael Mountain, Best Friends Animal Society"

"Oh, yeah—I think I've heard of that. Out in the middle of nowhere, right? In the desert."

"Yes. Kanab, Utah."

We order a couple of beers.

I asked him about Best Friends. He says it was founded twenty-five years ago by a ragtag band of animal lovers who dreamed of a place where homeless dogs and cats would never be euthanized. He tells me that it has grown to a $35 million operation (the same size as PETA); that Best Friends rescued 6,000 animals during Hurricane Katrina; and that they are no longer Best Friends Animal *Sanctuary* but have re-organized as Best Friends Animal *Society*—a nation-wide network of people and grass-roots community organizations devoted to saving animals.

I am impressed. But I am more impressed when our discussion turns to the strange ways people think about animals. He gets it: Human attitudes toward other species are inevitably paradoxical and inconsistent. He confesses to some of his personal moral lapses. He is a vegan and does not eat any animal products. But he purchases pigs' ears for his dogs to chew on. The dogs love them but Michael can't stop thinking about the poor pigs.

Then he tells me about the convoluted ethical guidelines that he follows in dealing with the biting horseflies that buzz around his house in the summer.

"Here is my rule," he says. "If I am walking outside and a horsefly bites me, it is permissible for me to swat it, just like you would a mosquito. However, if the horsefly comes into my house, I have to rescue it and take it outside." Smiling, he adds, "Where it will bite me the next time I go for a walk."

"Huh? That's completely ass-backward," I say. "It should be OK to kill a fly that has invaded your house, your home territory, and not OK for you to kill one outdoors when you are on its home territory? Is there a rationale for your rule?"

He laughs.

"Of course. There is always a rationale. But a rationale is not necessarily rational. I suppose my rule is much like the philosophy behind Best Friends. You can't save all the animals in the world, but the ones that come into your care, you are responsible for. So, once the horsefly enters my house, I have a responsibility to treat it with kindness."

Here is a rare bird: a morally serious person who can laugh at himself. But the guy seems like the real deal.

I glance at my watch. We have been talking for two hours. It's closing in on midnight and we are the only ones left in the place. As we get up to leave, Michael says that I should come out to Kanab—to talk more and see what Best Friends is about.

"Maybe I'll take you up on it," I say.

I get back to my hotel room and telephone Mary Jean. "Would you be interested in going out to Utah next summer?"

AN ANIMAL SANCTUARY IN THE MIDDLE OF NOWHERE

Mary Jean and I get off the plane in Las Vegas, pick up a black Hyundai from Avis, and head north up I-15. To get to Kanab, Utah, you drive two hours to Saint George, turn off the interstate, and drive another couple of hours on two-lane blacktops, dipping into Arizona through Colorado City (usually referred to in the media as a "polygamy enclave") and the Kaibab Paiute Indian Reservation, and on into Kanab, a town with one traffic light, and 3,769 full-time (human) residents.

The next morning, we head out to Best Friends which is five miles out of town on Highway 89. I expect the sanctuary to look like a big petting zoo. Wrong.

Nestled next to the Delaware-sized Grand Staircase-Escalante National Monument, the 3,700-acre sanctuary is encompassed within another 30,000 acres the organization leases from the Bureau of Land Management. The vista is of biblical proportions—big sky and miles of sandstone cliffs and mesas that remind me of the names of the colors in the giant box of Crayola crayons my sister and I used to fight over when we were little: Brick Red, Burnt Sienna, Mahogany, Raw Umber, Carnation Pink, Copper, Maroon. I feel as if I've seen this place before. Later I learn that I have. Many of my favorite TV shows as a child were shot in or around Angel Canyon—*The Lone Ranger, The Adventures of Rin Tin Tin, Lassie, Have Gun—Will Travel, Gunsmoke*—even *Death Valley Days* starring Ronald Reagan. In addition, since the 1920s, nearly one hundred feature films have been shot here including *Planet of the Apes, The Greatest Story Ever Told, The Man Who Loved Cat Dancing,* and *The Outlaw Josey Wales.*

Best Friends offers daily tours for the 30,000 visitors that stop by every year, but Michael has arranged a special behind-the-scenes tour for us. Our guide is Faith Maloney, a cheery sixty-five-year-old Englishwoman who seems to know the names of every one of the 1,700 rescued dogs, cats, pigs, horses, rabbits, donkeys, peacocks, guinea pigs, and parrots who reside in the sanctuary. Michael and Faith were part of a small group referred to reverently as "the founders." Best Friends grew out of the idealistic vision of some young people who got together in the mid-1960s with a desire to do some good in the world. ("We were not hippies," Michael warns me. "If anything, we were the anti-hippies. For example, our rule was no drugs.") After a stint in the Yucatan, the group became involved in politics, religion, and social services before disbanding, but some of them got back together in the late 1970s and discovered they had a common interest in saving animals.

In the early '80s, the group stumbled on Kanab Canyon, which they renamed Angel Canyon. Despite its remoteness, they decided that the canyon was just the place to establish a home for animals no one else wanted. They did not foresee that their small shelter in southwestern

Utah would become one of the nation's largest animal protection organizations, that an army of volunteers would one day walk dogs, hose down pot-bellied pigs, and shovel horse manure; that the sanctuary would be the subject of a hit television series (National Geographic's *Dogtown*); or that Best Friends would play a pivotal role in the most successful campaign in the history of the American animal protection movement—the establishment of community-based spay/neuter and adoption programs that have already reduced the number of dogs and cats killed each year in animal shelters from fifteen million to fewer than two million.

We meet Faith at the visitor center and walk over to Piggy Paradise (Best Friends has a penchant for cute names). We see a volunteer from Virginia exercising a pot-bellied pig by tempting it with pieces of no-fat popcorn, and we talk to a farrier who is rebuilding the shattered hoof of an enormous draft horse that had been abandoned at a dump. Then we jump into Faith's car and take off to Casa del Calmar. It is a house—a real house (no animals are kept in cages at Best Friends)—for cats with incurable conditions like feline leukemia. The place is spotless, nary a whiff of cat pee even though purring cats are draped everywhere. Faith explains that the focus of Best Friends is on animals with special needs—a three-legged cat, a dog with a lump on its throat the size of a baseball, an eagle with a broken wing. Most of them come from other shelters that can't give them the longer-term care they need. This is the shelter of last resort for the blind, the deaf, the handicapped, the psychologically damaged.

After brief stops at Bunny House, Horse Haven, Parrot Garden, and Wild Friends (a rehab center for turtles, owls, hawks, bobcats, and songbirds), we turn into Dogtown Heights, a ninety-acre complex that is home to some 400 dogs. At Old Friends, a facility for aging dogs, I meet Ruby Benjamin, an energetic seventy-eight-year-old psychotherapist from Manhattan who volunteers for a couple of weeks at Best Friends every year. "My heart is here," she tells me. "When I come to Best Friends, it is like getting a big hug."

In the main building at Dogtown, we are introduced to Cherry, a smallish black and white pit bull placidly lying on a cushion under the desk of a young woman typing on a computer. The dole-eyed Cherry looks completely laid back—you would never guess that she was one of nearly two dozen of quarterback Michael Vick's fighting dogs that were taken to Best Friends in the aftermath of the raid on his Bad Newz Kennels in rural Virginia.

Then it's a quick tour of the clinic. A visiting intern from a California vet school is spaying a cat in the surgical theater under the supervision of one of the six staff veterinarians. We stick our heads in the hydrotherapy room, then a technician proudly shows us his new state-of-the-art computerized X-ray machine. The clinic is better equipped than some of the hospitals I've been in.

Faith has arranged for us to have lunch with Frank McMillan. Dr. Frank, as he is known at Best Friends, is a veterinarian and an authority on the mental health of companion animals. He is the person in charge of rehabilitating Vick's pit bulls. He says that these animals are suffering from a canine version of post-traumatic stress disorder. Horribly mistreated at the hands of the dogfighters, their primary symptom is not aggression but fear. Frank and his team of animal behaviorists have been working with them for a year and a half. He originally expected the worst, but Dr. Frank has been pleasantly surprised. Twenty-one of the twenty-two dogs are making real progress. Several have passed the Canine Good Citizen test. And, so far, two have been placed in homes and one is in foster care.

The next morning, I return to spend the day as a volunteer. I am assigned to Dogtown. I am greeted by a man named Don Bain, a retired banker from Texas. He and his wife stumbled on Best Friends, fell in love with the place, and wound up buying a house in Kanab. Now he works in Dogtown as the "puppy socialization coordinator," which, for my money, is the world's greatest job title. He assigns me to work with a staff member named Terry. My job is to help her feed a dozen or so dogs. One of them is Shadow, one of the Vick pit bulls. He takes one look at me and snarls. Terry tells me that he is like that sometimes with strange men.

That's OK. I know that it is not his fault. He is a victim. At most animal shelters, he would have gotten the blue needle long ago. At Best Friends, he has a home for life, even if he turns out to be a jerk.

Then it's walk time.

Terry teams me up with another volunteer, Dora, a woman from Kansas City who works at Home Depot. She is on her way home from San Francisco and has driven two days so she could spend some time volunteering at Best Friends. Dora snaps a leash on a brown Lab-ish looking dog named Cinderella. I get Lola, a mixed-breed I fall in love with because she looks to be the identical twin of Frisky, my childhood dog. Off we go on the walking trail through the pinion pines, juniper, rabbit bush, and prickly pear, the White Cliffs of the Grand Staircase in the distance. The dogs are happy and so are we . . . at least until Dora and I, busily talking about dogs, get hopelessly lost. Terry finds us just as the wind kicks up, the temperature drops, thunder rolls, and the rain starts to fall.

Over dinner, we tell Michael about our experiences at the sanctuary. After hanging around Kanab for a week, Mary Jean and I are still astonished at what a handful of dreamers with a vision have accomplished in the Utah desert. It is a first-class operation. The bathrooms are clean, and the staff returns your phone calls. More impressively, all the animals are individuals at Best Friends. The staff does not talk about "dogs" or "cats" or "horses"; they talk about James, Minda, and Moonshine (a black and white guinea pig). There is an eerie tranquility about the place. Everyone is *so* nice. It's all a little spooky.

The philosopher Immanuel Kant argued that humans should not mistreat animals, but only because he felt that animal cruelty makes people violent toward other people. Michael tells me the vision of the founders of Best Friends was the converse of Kant's dictum; they were convinced that being nice to animals can make us kinder to other humans.

The scope of his vision tires me out. Michael thinks differently than I do. He thinks BIG. Too big for me.

I change the subject. "Have you always been good with animals?"

He surprises me. "Well, I don't get gooey over them. I am better at organizing and editing and putting people together than I am with taking care of animals."

But then he tells me about the ants in his kitchen. "They are cute, the ants. They are basically like a sanitation department. The ants come into the kitchen on these patrols. If they don't find anything, they go back outside. But if Miss Popsicle, my cat, has left a little piece of food somewhere, they mount a military operation to transport it outside. It is quite amazing. It takes a lot of ants to carry a piece of cat food. When I see them putting this much work into it, I try to help them out by getting them and the piece of cat food on a piece of tissue paper and take them outside."

I try to imagine a man who can chat up the CEO of a Fortune 500 corporation, who is on a first-name basis with Hollywood's A-list, but who claims he is not particularly good with animals, down on hands and knees on his kitchen floor gently brushing ants so small you can hardly see them onto a Kleenex and transporting them back to their nest, just to make their day a little easier.

THE SEA TURTLE LADY

Michael Mountain is a dreamer trying to pull off a kindness revolution. But most animal people are more like Judy Muzzy. She owned the Beach Combers Hair and Nail Studio, a beauty shop on Edisto Island, South Carolina, and she spent her spare time trying to save endangered sea turtles.

The path to becoming an animal rescuer usually begins with a concern for a single animal, an emaciated stray dog perhaps or a cat on the euthanasia list at an animal shelter with a three-day-and-out policy. Sea turtle people are different. They don't get much back in the way of the warm and fuzzies. Indeed, many of them will never even see one of the animals they are trying to save. It is enough for them to know that, when they patrol the beach at dawn, a 300-pound female loggerhead is waiting just beyond the surf for the sun to set so she can lumber ashore, laboriously dig a nest two feet deep in the sand, and lay a hundred squishy eggs the size of Ping-Pong balls which will hatch a couple of months later, and that, with luck, one in a thousand hatchlings will survive to repeat the process twenty-five years later.

For a beach town, Edisto is backwater. It has no motel, no putt-putt, no McDonald's, and no water park. There is, however, a grungy Bi-Lo supermarket and Whaley's, a slightly seedy bar with a pool table, good food, and an easy ambiance two blocks off the main road near the town water tower. Mary Jean and I were in Whaley's one Sunday evening sipping beer at the bar while waiting for our shrimp sandwiches. Mary Jean was talking to the woman sitting on the barstool next to her. My attention was divided between the NASCAR race on the television above the bar and two guys across from me who were talking fishing and drinking oyster shooters for dinner. (You drop a raw oyster in the bottom of a shot glass, add an ounce of Smirnoff, a little Tabasco sauce, and a squeeze of lemon. Toss it down in one gulp and chase it with a swig of Bud Lite.)

I had watched them work their way through a dozen oysters when I heard Mary Jean say to her new friend, "Oh, you need to talk to my husband. He studies people who love animals." Judy, it turns out, was nuts over loggerhead turtles—the endangered lumbering giants that nest on beaches from Texas to North Carolina. Mary Jean and I switched seats.

Judy told me she had moved to Edisto ten years before after her marriage broke up. She worked two jobs for a couple of years and saved enough money to open her hair salon. I asked her about sea turtles, and she lit up, whipped out her cell phone, and started showing me pictures of giant turtle tracks on the beach, opened nest cavities, and newly hatched button-cute babies. Judy was part of a team of volunteers that patrols the beach at dawn, recording nest locations and crawls (the three-foot wide trails in the sand that nesting females leave on the beach). Sometimes they dig up vulnerable nests and move the eggs to safer ground. Once the eggs begin to hatch, she returns and records their fate—the numbers of successfully hatched eggs and dead babies.

Judy invited me to join her the next day on the dawn patrol but we were leaving town. I promised to come back.

■ ■ ■

A year later, I am sitting in Judy's living room drinking sweet tea. It is dark and cool, which is good because it is nearly one hundred degrees outside with 98% humidity. She introduces me to her arthritic chocolate Lab named OB (because he only has one ball) and to Megan, her eighteen-year-old granddaughter who also helps with the turtle rescue project. They fill me in on the basics of loggerhead reproductive biology. Female turtles spend their lives roaming the oceans, only coming on land every two or three years to lay their eggs. The males never come ashore. A female will typically produce several clutches in a season, which at Botany Bay averages about sixty eggs per nest. The nests are architectural marvels. They are shaped like two-foot-deep chemistry flasks with the egg cavity mushrooming out of the bottom of a narrow tunnel. The female excavates the nest with her rear flippers, and then, once the eggs are deposited, she refills the nest hole and obscures its location from predators.

Many eggs never hatch. A raccoon will dig up the nest or a ghost crab will hide out in the egg cavity, gorging on yolks and embryos. In about fifty days, the nestlings will pip their shells, spend a couple of days resting in the nest cavity and absorbing egg yolk, and then start digging their way to the surface. They almost always emerge at night, and instinctively know to make their way toward the surf, drawn by the open sky over the ocean and the reflection of moonlight on the water.

Loggerheads are an endangered species. Even under the best of circumstances, only one-tenth of one percent of the hatchlings will reach reproductive age and return back to these beaches to continue the cycle. Everything eats baby turtles. But sea turtles are not endangered because of raccoons or ghost crabs or sharks or birds. No, they may go the way of the passenger pigeon because they swallow chunks of Styrofoam floating in the open ocean or plastic shopping bags which they mistake for jellyfish or they get caught in commercial fishing nets or poisoned by oil spills. Then there is the loss of nesting habitat because of beachfront development. At Edisto Island, light pollution is a big problem. Instead of heading toward the surf, hatchlings can be drawn inland by the lights from condos or the gas station just across the road.

The South Carolina Department of Natural Resources runs the sea turtle protection program, but it is made possible by 800 volunteers like Judy who monitor virtually every mile of turtle beach in the state during the breeding season. The program has several purposes. One is scientific. The data gathered by the volunteers is invaluable. Thanks to the efforts of the beach patrollers, biologists know the exact location of nearly every sea turtle nest in the state. They can tell you for any given stretch of beach, the number of crawls, the number of nests, how many eggs were laid, the proportion of dead and live hatchings, and the sources of their mortality. In 2020, for example, the turtle patrol located 5,559 nests on South Carolina beaches, 249,304 eggs hatched, with a successful hatch rate of 44.8%. The average incubation time was fifty-three days, and the typical nest included 114 eggs.

The second purpose of the sea turtle protection program is to increase the percentage of hatchlings that survive. Volunteers are trained and certified. When they find a nest that is too close to the water line or not buried deep enough, they are authorized to relocate it to safer ground. Very carefully, a handful of sand at a time, they dig down into the nest cavity. Then, keeping each egg up-side up, they put it in a bucket, dig the new nest in a better location, and place the eggs in the new nest in the exact order in which they were extracted. On beaches where nests are prone to predation, volunteers sometimes "cage" them by staking a protective fence over nests. Once the eggs have hatched, the volunteers collect additional data by digging open every fourth nest. They record the number of hatched and unhatched eggs, and the number that hatched but died in the nest. In the dog days of August, this is hot, dirty, smelly, yolky work. One day, Judy dug open a nest and found twenty dead hatchlings. It was a heartbreaker.

But every now and then, at the bottom of a nest, she will come across a baby turtle that is still alive but who, for some reason, could not make it to the surface on its own.

And she saves it.

Judy has been a volunteer in South Carolina's sea turtle nest protection program for five years. I ask her why she does it.

"Well, at first it was the thrill of riding a four-wheeler down the beach at dawn, and it was exciting coming on those big turtle tracks—the crawls. But the first time you dig up a hatched-out nest to count the eggs and you see one of those little babies alive that did not make it to the surface, it just melts your heart."

How many turtles do you think you have saved over the last five years? I ask.

"Oh, lots, maybe hundreds." Then wistfully, she adds, "But I know that, realistically, only one in a thousand will survive."

Then she laughs. "But, in my mind, every baby turtle that I ever helped makes it. I don't know what happens to the rest of them. But I know mine make it."

Then we make plans to meet the next morning for the dawn patrol. Judy says to bring water and bug spray.

■ ■ ■

Six A.M. Blue-and-gold Maxfield Parrish sky.

The team consists of Judy, myself, a woman named Sherri Johnson, and April Fludd, a seventh-grader who has been working on the sea turtle project for two summers now. Our section of beach is on Botany Bay Plantation Heritage Preserve—6,500 acres of sand, salt marsh, old fields, and live oak forests. It is a jewel—one of the most pristine beaches on the Atlantic Coast.

We pick up a couple of ATVs loaded with gear: several five-foot-long T-shaped probes you use to locate the nests, marker flags, a couple of rolled-up mesh cages to keep the raccoons out, a GPS, a big bucket in case we have to relocate a nest, bright orange signs warning beach walkers if they mess with a turtle nest, bad things will happen.

Judy tells me to hop on the back of the red Honda four-wheeler with her, and April gets on the green Yamaha with Sherri. We are off. The sun is coming up over the marsh. The air has a little chill, and we hear only birds. We zoom down the half-mile trail toward the ocean, across the salt marsh. An osprey is cruising overhead, a family of wood storks forages in the shallows.

Judy looks back and yells to me over the rumble of the ATV, "See why I come out here. This is my church." She is right. Dawn at this beach is as transcendent as sitting in the nave at Chartres Cathedral staring at stained glass.

After making our way through a very buggy bog, we hit the sand and start cruising, looking for crawls. Other than a shrimp boat trawling 500 yards offshore and a flock of low-flying pelicans in tight formation, the beach is empty. The ocean is placid, like a farm pond, and the coast of Africa looks to be just over the horizon.

Bad luck. We race the four-wheelers down to where the North Edisto River separates Botany Bay Island from Seabrook Island, a tony, gated golf-and-tennis community for wealthy retirees, but we don't see a crawl. Zip. Nada. We turn around empty handed. I am discouraged.

Then, good luck. On the way back, we run into Chris Salmonsen, a wildlife biologist who is responsible for the section of beach next to Judy's. We start to talk. When I tell him I am interested in turtle volunteers, people like Judy and Sherri and April, he smiles and says he could not do his job without them. We agree to meet later when he will fill me in on human-loggerhead relationships.

Chris, forty-six, has a background in environmental education and has worked with turtle volunteers from Texas to South Carolina.

"Tell me about the turtle people."

Some of them, he says, usually the men, like racing ATVs up and down the beach. He gives me the impression that they usually don't last very long. Most of the serious volunteers are women.

He says, "A lot of the volunteers have a space in their lives that they need to fill, and the turtles help them fill that space. It is like going to church on Sunday. It is their religion."

I mention that Judy had told me exactly the same thing.

Judy is different from some of the volunteers, Chris says. She has lots of things going on in her life. Saving turtles is just one of the things she does. She also runs a business and is an artist.

Chris goes on. "That's not true of all the volunteers. Some of them are completely obsessed. They wear turtle shirts all the time, and their house is covered with turtle stuff. They want everyone to know they are turtle people. It *is* their identity. "

He tells me about the time he was having dinner with a woman who was a turtle rescuer in Texas. When he ordered the shrimp special, she broke out into tears. It turns out that the woman was in a fight with local shrimpers over the use of TEDs—turtle excluder devices—which enable loggerheads to escape from shrimp nets. Despite the use of TEDs, she felt that every shrimp cocktail translated to a dead loggerhead.

The next day I am up at dawn again, this time patrolling with Chris and his volunteers: a college student named Rosa who goes out five days a week, and Marie, a photographer who is carrying a Canon SLR with a telephoto lens as long as my arm. I hop on the back of Rosa's ATV and we take off through the bog next to the marsh. Today, the bugs in the marsh trail are really bad—*African Queen* bad. They bite through our clothes and we are constantly swatting at them with our baseball caps. Rosa has a trickle of blood running down her leg.

We drive down the beach and hit the first crawl after just a quarter mile. Rosa spots it first which means that she had to find the hole leading to the nest cavity. She takes one of the probes and goes to work. You use the probe to locate the narrow tunnel leading to the nest cavity. The sand in the tunnel is packed more loosely than the surrounding ground. Probing is a delicate process. You have to use enough pressure to sink the tip of the probe into the sand. But if you put too much weight on it and hit the loose sand of the nest hole, the probe will dive into the cavity and skewer some of the eggs. Probing takes a deft touch; roughly 10% of all egg losses are caused by probes.

It takes fifteen minutes of careful repeated probing around an area two feet in diameter before Rosa feels the loose sand of the tunnel. She hands me the probe. I push it into the sand next to the tunnel. It is not going anywhere. I move it over an inch, push it down again and, bingo, the loose sand in the tunnel immediately gives. Chris drops to his knees and starts to dig with one hand. By the time he reaches the eggs, his arm is engulfed in sand. He tells me to stick my arm in the hole, to feel the egg. It is leathery, like an alligator's egg, and somehow seems alive.

We fill the tunnel back up with sand, and Chris flags the nest, notes the GPS coordinates, and logs the data into his notebook. He will enter it into the computer when he gets back to the house, and it will show up online within twelve hours. We jump back on the ATVs and Chris quickly spots another crawl. He grabs the probe. We go through the drill and are off again. I am feeling a little giddy. We have found two nests. I've got the sea turtle fever.

Several weeks later, I caught up with Meg Hoyle who coordinates the volunteer program on Botany Bay. I wanted her take on what makes people like Judy, Rosa, Sherri, and April get up at dawn and fight off biting flies when it is pushing one hundred degrees in the shade and come home smelling like rotten eggs. And this is just to help animals that you hardly ever see and creatures that will almost certainly be munched up by a predator long before they are old enough to breed.

She said, "Sometimes I don't understand it myself. Some of the volunteers walk the beach morning after morning. And in an entire summer, they might come across twelve nests. Most days, they will never find a single crawl and many of them will never see a turtle all season. Yet they stick with it. They are looking for a connection with the natural world that we have gotten away from. We all need that connection with animals and the outdoors."

THE ANTHROZOOLOGY OF EVERYDAY LIFE

Meg is right. Most people do feel the need to connect with animals and nature. But we have this need to varying degrees. There are not many people like Michael Mountain, people who will not kill the ants invading their kitchen. There are, however, scads of people like Judy Muzzy, people who have jobs and families, people who are doing what they can in small ways to connect with animals, and are not particularly bothered by the many paradoxes in their interactions with other species. They don't agonize over whether one should throw a switch that would send a hypothetical train careening into an old man or a group of endangered chimpanzees. They don't care whether the correct route to animal liberation runs through Jeremy Bentham or Immanuel Kant. Nor do they feel guilty over the fact that they refuse to eat beef but wear leather shoes.

I have—mostly—come to accept my own hypocrisies. The yahoo within me says that it is better to let Tilly outside than keep her imprisoned in the house all day even though I know she occasionally kills a towhee or a chipmunk. The yahoo tells me that the exquisite taste of slow-cooked pit barbecue somehow justifies the death of the hog whose loin I am going to slather with a pepper-based dry rub.

Moral intuitions change, however, and sometimes the yahoo and I make new arrangements. I don't eat veal anymore, we buy local eggs produced by happy hens, and I am willing to pay more for a "free-range" chicken because I want to think it led a better life than a Cobb 500. And when a grizzled old rooster fighter recently asked me if I would like to go with him to a five-cock derby in Kentucky, I said no thanks.

When I first started studying human-animal interactions I was troubled by the flagrant moral inconsistencies I have described in these pages—vegetarians who sheepishly admitted to me they ate meat; cockfighters who proclaimed their love for their roosters; purebred dog enthusiasts whose desire to improve their breed has created generations of genetically defective animals; hoarders who caused untold suffering to the creatures living in filth they claim to have rescued. I have come to believe these sorts of contradictions are not anomalies or hypocrisies. Rather, they are inevitable. They are evidence of our humanity.

New York University's Kwame Anthony Appiah is sometimes asked what he does for a living. When he replies he is a philosopher, the next question is usually, "So, what's your philosophy?" His standard response is, "My philosophy is that everything is more complicated than you thought."

What the study of human-animal interactions reveals is that our attitudes, behaviors, and relationships with the animals in our lives—the ones we love, the ones we hate, and the ones we eat—are, likewise, more complicated than you thought.

ACKNOWLEDGMENTS

My deep gratitude to the many animal people, from activists to zoologists, who have let me peer into their worlds and answered my naïve questions about their connections with other species.

For many years, my scholarly home has been the International Society for Anthrozoology, an enthusiastic group of researchers who transcend traditional academic pigeonholes. I am especially indebted to my fellow researchers Arnie Arluke, John Bradshaw, Alan Beck, Ben and Lynette Hart, Mikel Delgado, Anthony Podberscek, Andrew Rowan, and James Serpell for their encouragement and their contributions to the study of human-animal relationships. I owe much to the ethologist Gordon Burghardt, my longtime mentor and friend. David Henderson and Chris Diehm helped clarify the intricacies of animal rights philosophies for me. Rob Bass and Gail Dean read most chapters, and this book is much better for their insight and critiques. When in need of literary inspiration, I turned to Harry Greene, Robert Sapolsky, Elmore Leonard, and Merle Haggard.

This edition of the book would not have been published except for the efforts of my agent Rachel Sussman. The editorial staff at Harper has been a dream team. Executive Editor Gail Winston was the perfect taskmaster. She reined me in when I needed it and became the little voice in my head that said, "Always remember that your readers are smart." The book is better for the skilled hand and fact-checking of copyeditor Amy Hawley. Publisher Jonathan Burnham immediately understood the book's message and had the insight to know that it needed an additional chapter.

I could not ask for better colleagues than the faculty of the psychology department at Western Carolina University. For twenty-five years, Bruce Henderson and David McCord read drafts of my papers and let me know when I made a wrong turn. Much of my research has been conducted in collaboration with graduate and undergraduate students at Western Carolina University. I hope they had as much fun as I did.

Writing a book can make you crazy, and I got by with a little help from my friends—actually lots of help. My longtime kayaking pals helped me keep things in perspective by reminding me when I needed to go with the flow. For nearly fifteen years, I recharged my batteries every Tuesday night playing old fiddle and swing tunes at Guadalupe's Restaurant in Sylva, North Carolina, with the fiddler extraordinaire Ian Moore and the shaggy bassist Adam Bigalow. Special thanks go to Mac Davis who I first met when we moved into a small farmhouse up Sugar Creek. He was across the road plowing his tobacco field with a mule. Forty years later, I am still running ideas by him, including many contained in this book.

My family was extraordinarily supportive during the months I spent buried in a dingy basement office amid piles of reprints and cups of cold coffee. My brother, sister, and mother were a source of constant encouragement. My collaborator-for-life is my wife, Mary Jean. Early in our relationship proved her mettle by helping me demonstrate that angry mother alligators would actually attack a human intruder to defend their young. More recently, she uncomplainingly read every sentence in the new edition of this book and has generally kept me sane. Our children, Adam, Katie, and Betsy, all fine writers, cheerfully critiqued chapters. I could always rely on them for honest advice like, "Dad, this sentence sucks."

Finally, a crunchy salmon treat to Tilly, who still spends many a drowsy afternoon lying in a rocking chair, keeping me company, and watching me write, occasionally meowing so I will rub her belly, reminding me why we bring animals into our lives.

NOTES

INTRODUCTION

xiii *tell a lot about who we are*: Marc Bekoff speech to the Farm Sanctuary Hoe Down (www.farmsanctuary.org) in Orland, California, on May 16, 2009.

xviii *fall victim to the claws of pet cats*: Loss, S. R., Will, T., & Marra, P. P. (2013). The impact of free-ranging domestic cats on wildlife of the United States. *Nature Communications*, 4(1), 1–8.

xviii *Kansas investigators informed a group of cat owners*: Stuchhury, B. (2007). *Silence of the songbirds*. Toronto: HarperCollins.

xx *bone in their penis*: The penis bone is called the *baculum* or *os penis*. Penis bones are found in many species of mammals including primates. Ramm, S. A. (2007). Sexual selection and genital evolution in mammals: A phylogenetic analysis of baculum length. *American Naturalist*, 169, 360–369.

xxii *four and a half million Americans are bitten by dogs*: Dixon, C. A., Pomerantz, W. J., Hart, K. W., Lindsell, C. J., & Mahabee-Gittens, E. M. (2013). An evaluation of a dog bite prevention intervention in the pediatric emergency department. *Journal of Trauma and Acute Care Surgery*, 75, 308–312.

xxii *canine train wreck*: Serpell, J. A. (2003). Anthropomorphism and anthropomorphic selection: Beyond the "cute response." *Society & Animals*, 11(1), 83–100.

xxiii *back issues of sleazy supermarket tabloids*: Herzog, H. A., & Galvin, S. L. (1992). Animals, archetypes, and popular culture: Tales from the tabloid press. *Anthrozoös*, 5, 77–92.

xxiv *the "troubled middle"*: Donnelley, S. (1989). Speculative philosophy, the troubled middle, and the ethics of animal experimentation. *Hastings Center Report*, 19, 15–21.

CHAPTER 1

1 *human qualities: arrogance and ignorance*: Flynn, C. (2008). *Social creatures: A human and animal studies reader*. Brooklyn, NY: Lantern Books (p. xiv).

2 *recent progress report*: McCardle, P. D., McCune, S., & Griffin, J. A. (Eds.) (2020). *Human-animal interaction research: A decade of progress*. Lausanne: Frontiers Media SA.

4 *his Chow Chow was in the room*: Coren, S. (2013, February 11). How therapy dogs almost never came to exist. (Blog post) https://www.psychology

today.com/us/blog/canine-corner/201302/how-therapy-dogs-almost-never-came-exist (Downloaded December 16, 2020).

5 *German researchers took this question on*: Brensing, K., Linke, K., & Todt, D. (2003). Can dolphins heal by ultrasound? *Journal of Theoretical Biology*, 225(1), 99–105.

6 *They found that all of them were methodologically flawed*: Marino, L., & Lilienfeld, S. O. (1998). Dolphin-assisted therapy: Flawed data, flawed conclusions. *Anthrozoös*, 11(4), 194–200. Marino, L., & Lilienfeld, S. O. (2007). Dolphin-assisted therapy: More flawed data and more flawed conclusions. *Anthrozoös*, 20, 239–249. Marino, L., & Lilienfeld, S. O. (2021). Third time's the charm or three strikes you're out? An updated review of the efficacy of dolphin-assisted therapy for autism and developmental disorders. *Journal of Clinical Psychology*, 1–15.

7 *suffered traumatic injuries*: Mazet, J. A., Hunt, T. D., & Zoccardi, M. H. (2004). *Assessment of the risk of zoonotic disease transmission to marine mammal workers and the public*. Final report. United States Marine Mammal Commission, RA No. K005486–01.

7 *forty-nine studies on the effectiveness of animal-assisted therapies*: Nimer, J., & Lundahl, B. (2007). Animal-assisted therapy: a meta-analysis. *Anthrozoös*, 20(3), 225–238.

7 *interventions involving animals have produced mixed results*: O'Haire, M. E., & Rodriguez, K. E. (2018). Preliminary efficacy of service dogs as a complementary treatment for posttraumatic stress disorder in military members and veterans. *Journal of Consulting and Clinical psychology*, 86(2), 179. Binfet, J. T. (2017). The effects of group-administered canine therapy on university students' wellbeing: A randomized controlled trial. *Anthrozoös*, 30(3), 397–414. McCullough, A., Ruehrdanz, A., Jenkins, M. A., Gilmer, M. J., Olson, J., Pawar, A., & Grossman, N. J. (2018). Measuring the effects of an animal-assisted intervention for pediatric oncology patients and their parents: A multisite randomized controlled trial. *Journal of Pediatric Oncology Nursing*, 35(3), 159–177. Mueller, M., Anderson, E. C., King, E. K., & Urry, H. L. (in press). Mechanisms of anxiety reduction in animal-assisted interventions for adolescents.

8 *nursing home residents*: Thodberg, K., Sørensen, L. U., Christensen, J. W., Poulsen, P. H., Houbak, B., Damgaard, V., & Videbech, P. B. (2016). Therapeutic effects of dog visits in nursing homes for the elderly. *Psychogeriatrics*, 16(5), 289–297.

9 *kids with autism could not be trusted*: Davis, T. N., Scalzo, R., Butler, E., Stauffer, M., Farah, Y. N., Perez, S., & Coviello, L. (2015). Animal assisted interventions for children with autism spectrum disorder: A systematic review. *Education and Training in Autism and Developmental Disabilities*, 316–329.

9 *most common problems with these studies*: Crossman, M., & Herzog, H. (2019). The research challenge: Threats to the validity of animal-assisted therapy studies and suggestions for improvement. In A. H. Fine (Ed.),

Handbook on animal-assisted therapy: Theoretical foundations and guidelines for applying animal-assisted interventions (fifth edition). New York, NY: Elsevier.

9 *one of my first research projects*: Herzog, H. A. (1993). The movement is my life: The psychology of animal rights activism. *Journal of Social Issues*, 49(1), 103–119.

10 *research by John Nezlek and his colleagues*: Nezlek, J. B., Cypryanska, M., & Forestell, C. A. (2020). Dietary similarity of friends and lovers: Vegetarianism, omnivorism, and personal relationships. *Journal of Social Psychology* (pp. 1–7). doi: 10.1080/00224545.2020.1867042.

14 *Sam and the anthrozoologist Anthony Podberscek*: Podberscek, A. L., & Gosling, S. D. (2005). Personality research on pets and their owners: Conceptual issues and review. In A. Podberscek, E. S. Paul, & J.A. Serpell (Eds.), *Companion animals and us: Exploring the relationships between people and pets* (pp. 143–167). Cambridge, UK: Cambridge University Press. Puskey, J. L., & Coy, A. E. (2020). Exploring the effects of pet preference, presence, and personality on depression symptoms. *Anthrozoös*, 33(5), 643–657. Reevy, G. M., & Delgado, M. M. (2015). Are emotionally attached companion animal caregivers conscientious and neurotic? Factors that affect the human–companion animal relationship. *Journal of Applied Animal Welfare Science*, 18(3), 239–258. Gosling, S. D., Sandy, C. J., & Potter, J. (2010). Personalities of self-identified "dog people" and "cat people." *Anthrozoös*, 23(3), 213–222.

15 *play out in our interactions with companion animals*: Westgarth, C., Brooke, M., & Christley, R. M. (2018). How many people have been bitten by dogs? A cross-sectional survey of prevalence, incidence and factors associated with dog bites in a UK community. *Journal of Epidemiology and Community Health*, 72(4), 331–336. Turcsán, B., Range, F., Virányi, Z., Miklósi, Á., & Kubinyi, E. (2012). Birds of a feather flock together? Perceived personality matching in owner–dog dyads. *Applied Animal Behaviour Science*, 140(3–4), 154–160. Dodman, N. H., Brown, D. C., & Serpell, J. A. (2018). Associations between owner personality and psychological status and the prevalence of canine behavior problems. *PLoS One*, 13(2), e0192846. Gobbo, E., & Zupan, M. (2020). Dogs' sociability, owners' neuroticism and attachment style to pets as predictors of dog aggression. *Animals*, 10(2), 315. Hauser, S. (2016). Like owner, like dog: Relationship between owner and dog personality. Thesis, Queensland University.

15 *the result of emotional contagion*: Schöberl, I., Wedl, M., Beetz, A., & Kotrschal, K. (2017). Psychobiological factors affecting cortisol variability in human-dog dyads. *PLoS One*, 12(2), e0170707. Sundman, A. S., Van Poucke, E., Holm, A. C. S., Faresjö, Å., Theodorsson, E., Jensen, P., & Roth, L. S. (2019). Long-term stress levels are synchronized in dogs and their owners. *Scientific Reports*, 9(1), 1–7.

16 *Some scientists believe the roots of cruelty lie in our evolutionary history*:

Nell, V. (2006). Cruelty's rewards: The gratifications of perpetrators and spectators. *Behavioral and Brain Sciences*, 29, 211–224.

16 *The anthropologist Margaret Mead wrote*: This quote is cited on page 80 of Lockwood, R., & Hodge, G. R. (1998). The tangled web of animal abuse: The links between cruelty to animals and human violence. In R. Lockwood & F. R. Ascione (Eds.), *Cruelty to animals and interpersonal violence* (pp. 77–82). West Lafayette, Indiana: Purdue University Press.

17 *Alan Felthous and Stephen Kellert*: Felthous, A. R., & Kellert, S. R. (1986). Violence against animals and people: Is aggression against living creatures generalized? *The Bulletin of the American Academy of Psychiatry and the Law*, 14(1), 55–69.

17 *they had abused animals in the past*: Marceau, J. (2019). Beyond cages: Animal law and criminal punishment. New York: Cambridge University Press.

18 *"I beat a puppy, I believe simply from enjoying the power"*: This quote is found in Blakemore, C. (2009, February 12). Darwin understood the need for animal tests. *The Times* (London). http://www.timesonline.co.uk/tol/comment/columnists/guest_contributors/article5711912.ece.

18 *presentations by Link advocates often begin with tales of tragedy*: See, for example, Lockwood, R., & Hodge, G. R. (1998). The tangled web of animal abuse: The links between cruelty to animals and human violence. In R. Lockwood & F. R. Ascione (Eds.), *Cruelty to animals and interpersonal violence* (pp. 77–82). West Lafayette, Indiana: Purdue University Press.

18 *did not have a known history of cruelty to animals*: The researchers in this study found that 21% of the 354 serial killers had a history of animal abuse. Most studies of male college students have also found animal abuse in the 20% to 30% range. The serial killer data is reported in Wright, J., & Hensley, C. (2003). From animal cruelty to serial murder: Applying the graduation hypothesis. *International Journal of Offender Therapy and Comparative Criminology*, 47, 71–88.

18 *animal cruelty is equally tenous*: Arluke, A., & Madfis, E. (2014). Animal abuse as a warning sign of school massacres: A critique and refinement. *Homicide studies*, 18(1), 7–22.

19 *called meta-analysis to combine the results of fifteen of these studies*: Patterson-Kane, E. (2016). The relation of animal maltreatment to aggression. In L. Levitt, G. Patronek, & T. Grisso (Eds.), *Animal maltreatment: Forensic mental health issues and evaluations* (pp. 140–158). Oxford University Press.

19 *A stronger version of Link thinking is called the violence graduation hypothesis*: This idea is sometimes referred to as the progression thesis. Beirne, P. (2004). From animal abuse to interhuman violence? A critical review of the progression thesis. *Society and Animals*, 12(1), 39–65.

20 *that pose a danger to the entire community*: O'Connor, J. (2020, January 10). A disturbing sign of psychopathy. *Washington Post*. https://www.washingtonpost.com/opinions/a-disturbing-sign-of-psychopathy/2020/01/10

/f21d0a6e-326d-11ea-971b-43bec3ff9860_story.html (Downloaded December 17, 2020).

20 *the backgrounds of 570 young adults*: Goodney-Lea, S. R. (2005). *Guns, explosives, and puppy dog tails: The social function of animal cruelty.* Unpublished doctoral dissertation, Indiana University.

20 *The students he interviewed*: Arluke, A. (2002). Animal abuse as dirty play. *Symbolic Interaction,* 25(4), 405–430.

21 *that they had abused animals*: Gupta, M. E. (2006). *Understanding the links between intimate partner violence and animal abuse: Prevalence, nature, and function.* Unpublished doctoral dissertation, University of Georgia, Athens, GA.

21 *have come to question*: Marceau, J. (2019). *Beyond cages: Animal law and criminal punishment.* Cambridge University Press. Piper, H., & Cordingley, D. (2009). The power and promulgation of the claimed links between human and animal abuse. *Power and education,* 1(3), 345–355. Taylor, N., & Signal, T. (2008) Throwing the baby out with the bathwater: Towards a sociology of the human-animal abuse 'Link'? *Sociological Research Online,* 13(1). http://www.socresonline.org.uk/13/1/2.html.

22 *"Animals are good to think with"*: Levi-Strauss, C. (1966). *The savage mind.* Chicago, IL: University of Chicago Press.

CHAPTER 2

23 *the dog than with the flea*: Greene, E.S. (1995). Ethnocategories, social intercourse, fear, and redemption: Comment on Laurent. *Society and Animals,* 3(1), 79–88.

23 *North Carolina, had a problem*: Cohen, R. (2009, July 19). The ethicist: Nesting blues. *New York Times.*

24 *in the Nobel Prize-winning psychologist Daniel Kahneman's book*: Kahneman, D. (2011). *Thinking, fast and slow.* New York: Macmillan.

25 *He called this trait "biophilia"*: Wilson, E. O. (1984). *Biophilia.* Cambridge, MA: Harvard University Press.

25 *easily we are manipulated by baby releasers*: Lutts, R. H. (1992). The trouble with Bambi: Walt Disney's Bambi and the American vision of nature. *Forest and Conservation History,* 36, 160–171. Cartmill, M. (1993). *A view to death in the morning.* Cambridge, MA: Harvard University Press.

26 *the late Harvard biologist who traced Mickey's evolution, said it best*: Gould, S. J. (1979). Mickey Mouse meets Konrad Lorenz. *Natural History,* 88(5), 30–36.

26 *canine versions of our neuroses*: Serpell, J. A. (2019). How happy is your pet? The problem of subjectivity in the assessment of companion animal welfare. *Animal Welfare,* 28(1), 57–66.

27 *In a Gallup poll*: Brewer, G. (2001). Snakes top list of Americans' fears. Gallup News Service. https://news.gallup.com/poll/1891/snakes-top-list-americans-fears.aspx (Downloaded December 20, 2020).

29 *a rattlesnake bite on the end of his tongue*: Gerkin, R., Sergent, K. C., Curry, S. C., Vance, M., Nielsen, D. R., & Kazan, A. (1987). Life-threatening airway obstruction from rattlesnake bite to the tongue. *Annals of Emergency Medicine*, 16(7), 813–816.

29 *six people in the United States die annually from a snake bite*: Forrester, J. A., Weiser, T. G., & Forrester, J. D. (2018). An update on fatalities due to venomous and nonvenomous animals in the United States (2008–2015). *Wilderness & Environmental medicine*, 29(1), 36–44.

29 *nurture in the development of snake fears*: For research on the origins of snake fears see Ohman, A., & Mineka, S. (2003). The malicious serpent: Snakes as a prototypical stimulus for an evolved module of fear. *Current Directions in Psychological Science*, 12(1), 5–9. LoBue, V., & DeLoache, J. S. (2008). Detecting the snake in the grass: Attention to fear-relevant stimuli by adults and young children. *Psychological Science*, 19(3), 284–289. DeLoache, J. S., & LoBue, V. (2009). The narrow fellow in the grass: Human infants associate snakes and fear. *Developmental Science*, 12(1), 201–207. Diamond, J. (1993). New Guineans and their natural world. In S. R. Kellert & E. O. Wilson (Eds.), *The biophilia hypothesis* (pp. 251–271). Washington, D.C.: Island Press. Burghardt, G. N., Murphy, J. B., Chiszar, D., & Hutchins, M. (2009). Combating ophidophobia: Origins, treatment, education and conservation tools. In S. J. Mullin & R. A. Seigel (Eds.), *Snakes: Ecology and conservations*. Ithaca, NY: Comstock Publishing.

29 *At the Kyoto Primate Institute*: Burghardt, G., & Bowers, R. I. (2017). From instinct to behavior systems: An integrated approach to ethological psychology. *APA Handbook of Comparative Psychology: Vol. 1. Basic Concepts, Methods, Neural Substrate, and Behavior*, J. Call (Ed.). Washington, D.C.: American Psychological Association. Matsuzawa, T., Hockings, K. J., Humle, T., & Carvalho, S. (2012). Chimpanzee interactions with nonhuman species in an anthropogenic habitat. *Behaviour*, 149(3–4), 299–324.

30 *"a perceptual bias for snakiness"*: LoBue, V., & Adolph, K. E. (2019). Fear in infancy: Lessons from snakes, spiders, heights, and strangers. *Developmental psychology*, 55(9), 1889–1907.

30 *"Biophilia," he wrote, "is not a single instinct"*: Wilson, E. O. (1993). Biophilia and the conservation ethic. In S. R. Kellert & E. O. Wilson (Eds.), *The biophilia hypothesis*. Washington, D.C.: Island Press. (p. 31).

32 *consumer attitudes toward flesh derived from stem cells*: Bryant, C. J., & Barnett, J. C. (2019). What's in a name? Consumer perceptions of in vitro meat under different names. *Appetite*, 137, 104–113.

34 *categorizing animals starts young*: Grief, M. L., Nelson, D. G. K., Keil, F. C., & Gutierrez, T. (2006). What do children want to know about animals and artifacts? *Psychological Science*, 17, 455–459.

34 *suggest that parts of the human brain may have evolved*: The idea of domain-specific knowledge is controversial. For contrasting views see Caramazza, A., & Shelton, J. R. (1998). Domain-specific knowledge system in the brain: The animate-inanimate distinction. *Journal of Cognitive Neuroscience*, 10,

1–34; and Chouinard, P. A., & Goodale, M. A. (2010). Category-specific neural processing for naming pictures of animals and naming pictures of tools: an ALE meta-analysis. *Neuropsychologia*, 48(2), 409–418. Gerlach, C., & Gainotti, G. (2016). Gender differences in category-specificity do not reflect innate dispositions. *Cortex*, 85, 46–53.

34 *the categories the rest of us put them in*: Arluke, A., & Sanders, C. (1996). *Regarding animals*. Philadelphia, PA: Temple University Press.

35 *moral concerns about the treatment of twenty species*: Leite, A. C., Dhont, K., & Hodson, G. (2019). Longitudinal effects of human supremacy beliefs and vegetarianism threat on moral exclusion (vs. inclusion) of animals. *European Journal of Social Psychology*, 49(1), 179–189.

35 *In Japan, attitudes toward creepy crawlies are more complex.*: Laurent, E. L. (2000). Children, "insects" and play in Japan. *Companion animals and us: Exploring the relationships between people and pets*, 61–89.

36 *anthrozoologist James Serpell*: Serpell, J. A. (2004). Factors influencing human attitudes to animals and their welfare. *Animal Welfare, 13*, S145–151.

37 *pigeons have shifted*: Jerolmack, C. (2008). How pigeons became rats: The cultural-spatial logic of problem animals. *Social Problems*, 55, 72–94.

38 *"emotionally satisfying yet thoroughly irrational"*: No author. (1977, September 12). Kill the crocodile. *New York Times*, p. 32.

39 *method to investigate how people make decisions about the use of animals in research*: Galvin, S. L., & Herzog, H. A. (1992). The ethical judgment of animal research. *Ethics & Behavior*, 2(4), 263–286.

40 *Haidt's theory of morality*: Haidt, J. (2001). The emotional dog and its rational tail: A social intuitionist approach to moral judgment. *Psychological Review*, 108(4), 814–834.

41 *Paul Rozin calls disgust the moral emotion*: Rozin, P., Haidt, J., & McCauley, C. R. (1999). Disgust: The body and soul emotion. In T. Dalgleish and M. Power (Eds.), *Handbook of cognition and emotion*. Chichester, UK: Wiley (pp. 429–445).

42 *Moral Machine Experiment*: Awad, E., Dsouza, S., Kim, R., Schulz, J., Henrich, J., Shariff, A., & Rahwan, I. (2018). The moral machine experiment. *Nature*, 563(7729), 59–64.

43 *the impersonal version (throwing the switch) does not*: Greene, J., & Haidt, J. (2002). How (and where) does moral judgment work? *Trends in Cognitive Sciences*, 6(12), 517–523.

43 *Here are two of Petrinovich's*: Petrinovich, L., O'Neill, P., & Jorgensen, M. (1993). An empirical study of moral intuitions: Toward an evolutionary ethics. *Journal of Personality and Social Psychology*, 64(3), 467–478.

44 *Researchers at Georgia Regents University*: Topolski, R., Weaver, J. N., Martin, Z., & McCoy, J. (2013). Choosing between the emotional dog and the rational pal: A moral dilemma with a tail. *Anthrozoös*, 26(2), 253–263.

44 *Answer these questions*: The bat and ball question still drives me crazy. Think of it this way: If the bat cost a $1.00 and the ball cost ten cents, then

the bat costs ninety cents more than the ball, not a dollar more. However, if the bat costs $1.05 and the ball costs five cents, then the bat costs a dollar more than the ball, and together they add up to $1.10. These examples are from Plous, S. (1993). *The psychology of judgment and decision making.* New York, NY: McGraw Hill.

45 *Our predilection for senseless revenge*: Sunstein, C. R. (2002 October). Hazardous heuristics. *University of Chicago Law & Economics Olin Law and Economics Working Paper No. 165* and *University of Chicago Public Law Research Paper No. 33.*

45 *heuristics is called framing*: Halliman, J. T. (2009). *Why we make mistakes.* New York, NY: Broadway Books. Marcus, G. (2008). *Kluge: The haphazard construction of the human mind.* New York: Houghton Mifflin. Kahneman, D., & Frederick, S. (2002). Representativeness revisited: Attribute substitution in intuitive judgment. In T. Gilovick, D. Grifin, & D. Kahneman (Eds.), *Heuristics and biases: The psychology of intuitive judgment* (pp. 49–81). Cambridge, UK: Cambridge University Press.

45 *the Nazi animal protection movement*: Arluke, A., & Sax, B. (1992). Understanding Nazi animal protection and the Holocaust. *Anthrozoös, 5*(1), 6–31.

47 *don't relish the idea that Adolf Hitler was a fellow traveler*: Berry, R. (2004). *Hitler: Neither vegetarian nor animal lover.* Brooklyn, NY: Pythagorean Books.

47 *But the anthrozoologist Boria Sax*: Sax, B. (2000). *Animals in the Third Reich: Pets, scapegoats, and the Holocaust.* A&C Black.

49 *how children and adults responded to an AIBO compared to a real dog*: Melson, G. F., Kahn, P., Beck, A., Friedman, B., & Edwards, N. (2009). Robotic pets in human lives: Implications for the human-animal bond and for human relationships with personified technologies. *Journal of Social Issues, 65,* 545–567.

51 *Serpell eloquently lays out the moral issues*: Serpell, J. A. (1996). *In the company of animals: A study of human-animal relationships.* Cambridge, UK: Cambridge University Press. (p. 177).

51 *Researchers at the University of Portsmouth*: Morris, P. H. (2008). Secondary emotions in non-primate species? Behavioural reports and subjective claims by animal owners. *Cognition & Emotion, 22*(1), 3–20.

52 *As the psychologist Eric Greene*: Greene, E.S. (1995). Ethnocategories, social intercourse, fear, and redemption: Comment on Laurent. *Society and Animals, 3*(1), 79–88.

52 *species ranging from ants to chimpanzees*: Herzog, H. A., & Galvin, S. (1997). Anthropomorphism, common sense, and animal awareness. In Mitchell, R. W., & Thompson, N. S. (Eds.). *Anthropomorphism, anecdotes, and animals.* SUNY Press (pp. 237–253).

52 *comparable to dogs and even primates*: Marino, L., & Colvin, C. M. (2015). Thinking pigs: A comparative review of cognition, emotion, and personality in *Sus domesticus. International Journal of Comparative Psychology, 28,*1–22.

53 *specific regions in the human brain*: Cullen, H., Kanai, R., Bahrami, B., & Rees, G. (2014). Individual differences in anthropomorphic attributions and human brain structure. *Social Cognitive and Affective Neuroscience*, 9(9), 1276–1280. Spunt, R. P., Ellsworth, E., & Adolphs, R. (2017). The neural basis of understanding the expression of the emotions in man and animals. *Social Cognitive and Affective Neuroscience*, 12(1), 95–105.

53 *beliefs about animal mental experiences*: Higgs, M. J., Bipin, S., & Cassaday, H. J. (2020). Man's best friends: attitudes towards the use of different kinds of animal depend on belief in different species' mental capacities and purpose of use. *Royal Society Open Science*, 7(2), 191162. Bastian, B., Loughnan, S., Haslam, N., & Radke, H. R. (2012). Don't mind meat? The denial of mind to animals used for human consumption. *Personality and Social Psychology Bulletin*, 38(2), 247–256.

53 *eating animals makes people more likely to deny*: Bastian, B., Loughnan, S., Haslam, N., & Radke, H. R. (2012). Don't mind meat? The denial of mind to animals used for human consumption. *Personality and Social Psychology Bulletin*, 38(2), 247–256.

53 *"What Is It Like to Be a Bat?"*: Nagel, T. (1974). What is it like to be a bat? *Philosophical Review*, 4, 435–450.

54 *what the psychologist Gordon Burghardt calls critical anthropomorphism*: Burghardt, G. M. (1991). Cognitive ethology and critical anthropomorphism: A snake with two heads and hognose snakes that play dead. In C. A. Ristau, (Ed.), *Cognitive ethology: The minds of other animals* (pp. 53–90). Hillsdale, NJ: Lawrence Erlbaum Associates.

CHAPTER 3

57 *lies at the heart of pet-keeping*: Pierce, J. (2016). *Run, Spot, run: The ethics of keeping pets*. Chicago: University of Chicago Press.

57 *whether we call them companions or pets*: Irvine, L. (2004). Pampered or enslaved? The moral dilemmas of pets. *International Journal of Sociology and Social Policy*, 24, 5–17 (p. 14).

57 *Antoine, a young Frenchman*: Gueguen, N., & Ciccotti, S. (2008). Domestic dogs as facilitators in social interaction: An evaluation of helping and courtship behaviors. *Anthrozoös*, 21, 339–349.

63 *The historian Keith Thomas argues that pets are animals that are*: Thomas, K. (1983). *Man and the natural world: A history of modern sensibility*. New York, NY: Pantheon.

63 *I prefer James Serpell's*: Serpell, J. A. Pet-keeping and animal domestication: A reappraisal. In J. Clutton-Brock (Ed.), *The walking larder: Patterns of domestication, pastorialism and predation* (pp. 10–21). London: Unwin Hyman.

64 *most animals in American homes*: Grier, K. C. (2006). *Pets in America: A history*. Chapel Hill, NC: University of North Carolina Press.

65 *Gary Francione rejects the institution of pet-keeping*: Francione, G., & Charlton, A. E. (no date). The case against pets. *Aeon.* https://aeon.co/essays/why-keeping-a-pet-is-fundamentally-unethical (Downloaded January 25, 2021).

65 *imagine that their pet had contracted a serious illness*: Kirk, C. P. (2019). Dogs have masters, cats have staff: Consumers' psychological ownership and their economic valuation of pets. *Journal of Business Research*, 99, 306–318.

66 *whether we call them companions or pets*: Irvine, L. (2004). Pampered or enslaved? The moral dilemmas of pets. *International Journal of Sociology and Social Policy*, 24, 5–17. (p. (14.)

68 *nineteenth-century France as well*: Kete, K. (1995). *The beast in the boudoir: Petkeeping in nineteenth-century Paris.* Berkeley: University of California Press.

69 *You won't go far wrong*: Holbrook, M. B. (2008). Pets and people: Companions in commerce. *Journal of Business Research*, 61, 546–552.

70 *benefits they derived from their relationships*: Herzog, H., Kowalski, R., Burgner, M., & Dunegon, C. (2003, May). Are pets really friends? Perceived benefits of relationships with companion animals. Paper presented at the meeting of the American Psychological Society, Atlanta, GA.

70 *not particularly attached to their pets*: Johnson, T. P., Garrity, T. F., & Stallones, L. (1992). Psychometric evaluation of the Lexington Attachment to Pets Scale. *Anthrozoös*, 5, 160–175.

71 *a 2018 survey by the National Opinion Research Center*: Applebaum, J. W., Peek, C. W., & Zsembik, B. A. (2020). Examining US pet ownership using the General Social Survey. *The Social Science Journal*, 1–10.

71 *with each additional person in a family*: Poresky, R. H., & Daniels, A. M. (1998). Demographics of pet presence and attachment. *Anthrozoös*, 11, 236–241.

71 *Pet industry lobbyist Mark Cushing*: Helflin, M. (2020, September 8). U.S. doesn't have enough pets, pet owners, and pet inclusive destination yet. Pet Products News. https://www.petproductnews.com/blog/u-s-doesn-t-have-enough-pets-pet-owners-and-pet-inclusive-destinations-yet/article_cffdc602-f1d6-11ea-8b5a-3f004e15c642.html.

72 *"we are going to put it there"*: Feldman, S. (2015, October 15). This Pet World Insider Moment Wth An Insider [video]. YouTube. https://www.youtube.com/watch?v=q19C9uE_TsU.

72 *An economic analysis commissioned by HABRI*: Clower, T. L., & Neaves, T. (2015). The Health Care Cost Savings of Pet Ownership. Human Animal Bond Research Initiative. https://habri.org/assets/uploads/HABRI_Report_-_Healthcare_Cost_Savings_from_Pet_Ownership_.pdf (Downloaded January 19, 2020).

73 *a recent review in the journal* Applied Developmental Science: Serpell, J., McCune, S., Gee, N., & Griffin, J. A. (2017). Current challenges to re-

search on animal-assisted interventions. *Applied Developmental Science*, 21(3), 223–233.

73 *owning a pet made a difference in their survival rates*: Friedmann, E., Katcher, A. H., Lynch, J. J., & Thomas, S. A. (1980). Animal companions and one-year survival of patients after discharge from a coronary care unit. *Public Health Reports*, 95(4), 307–312. Friedmann later replicated these effects. Friedmann, E., & Thomas, S. (1995). Pet ownership, social support, and one-year survival after acute myocardial infarction in the cardiac arrhythmia suppression trial (CAST). *American Journal of Cardiology*, 76, 1213–1217.

74 *A study of over 10,000 Germans and Australians*: Headey, B., & Grabka, M. M. (2007). Pets and human health in Germany and Australia: National longitudinal results. *Social Indicators Research*, 80(2), 297–311.

74 *Chinese women who owned dogs*: Headey, B., Na, F., Zheng, R. (2008). Pet dogs benefit owner's health. A "natural experiment" in China. *Social Indicators Research*, 87(3), 481–493.

74 *"Why Most Published Research Findings Are False"*: Ioannidis, J. P. (2005). Why most published research findings are false. *PLoS medicine*, 2(8), e124.

74 *widely cited studies in psychology*: Yong, E. (2018, November 19). Psychology's replication crisis is running out of excuses. *The Atlantic*.

74 *could not replicate forty-seven of fifty-three landmark cancer papers*: Baker, M. (2016). Biotech giant publishes failures to confirm high-profile science. *Nature News*, 530(7589), 141.

75 *Internet health sites such as WebMD*: Robinson, K. M. (2017, December 4). WebMD website. How pets help manage depression. https://www.webmd.com/depression/features/pets-depression##1 (Downloaded December 28, 2020).

75 *thirty-three published studies*: Hezog, H. (2019, December 3). The sad truth about pet ownership and depression. *Animals and Us*. https://www.psychologytoday.com/us/blog/animals-and-us/201912/the-sad-truth-about-pet-ownership-and-depression.

77 *a scientific report from the American Heart Association*: Levine, G. N., Allen, K., Braun, L. T., Christian, H. E., Friedmann, E., Taubert, K. A., & Lange, R. A. (2013). Pet ownership and cardiovascular risk: a scientific statement from the American Heart Association. *Circulation*, 127(23), 2353–2363.

77 *pet owners were healthier than non-pet owners have not panned out*: Bradshaw, J. (2017). *The animals among us: How pets make us human*. New York: Basic Books. Herzog, H. (2011). The impact of pets on human health and psychological well-being: fact, fiction, or hypothesis? *Current Directions in Psychological Science*, 20(4), 236–239.

77 *A study of 21,000 people in Finland*: Koivusilta, L. K., & Ojanlatva, A. (2006). To have or not to have a pet for better health? *PLoS ONE*, 1, 1–9.

77 *Researchers at the Australian National University*: Parslow, R. A., Jorm, A. F., Christensen, H., Rodgers, B., & Jacomb, P. (2005). Pet ownership and

health in older adults: Findings from a survey of 2,551 community-based Australians aged 60–64. *Gerontology,* 51(1), 40–47.

77 *people in their neighborhoods*: Christian, H., Bauman, A., Epping, J. N., Levine, G. N., McCormack, G., Rhodes, R. E., & Westgarth, C. (2018). Encouraging dog walking for health promotion and disease prevention. *American Journal of Lifestyle Medicine,* 12(3), 233–243.

77 *researchers at Purdue University*: Richards, E.A. (2017). Randomized controlled theory based, email mediated dog-walking intervention. (HABRI Grant report) https://habri.org/grants/projects/randomized-controlled-theory-based -e-mail-mediated-walking-intervention-differences-between-dog-owners -and-non-dog-owners (Downloaded September 11, 2020).

78 *found no association between dog ownership and obesity*: Miyake, K., Kito, K., Kotemori, A., Sasaki, K., Yamamoto, J., Otagiri, Y., & Ishihara, J. (2020). Association between pet ownership and obesity: A systematic review and meta-analysis. *International Journal of Environmental Research and Public Health,* 17(10), 3498.

78 *review of ten studies*: Kramer, C. K., Mehmood, S., & Suen, R. S. (2019). Dog ownership and survival: A systematic review and meta-analysis. *Circulation: Cardiovascular Quality and Outcomes,* 12(10), e005554.

78 *Among 300,000 Swedes*: Mubanga, M., Byberg, L., Egenvall, A., Ingelsson, E., & Fall, T. (2019). Dog ownership and survival after a major cardiovascular event: a register-based prospective study. *Circulation: Cardiovascular Quality and Outcomes,* 12(10), e005342.

80 *Therapy dogs can acquire and spread MSRA*: Santaniello, A., Sansone, M., Fioretti, A., & Menna, L. F. (2020). Systematic review and meta-analysis of the occurrence of ESKAPE bacteria group in dogs, and the related zoonotic risk in animal-assisted therapy, and in animal-assisted activity in the health context. *International Journal of Environmental Research and Public Health,* 17(9), 3278.

80 *more emotional support from their dogs and cats than from their husbands or kids*: Serpell, J. (2003). Anthropomorphism and anthropomorphic selection—Beyond the "cute response." *Society & Animals,* 11(1), 83–100.

80 *find it harder to leave their dog for a week than their human partner*: Rover. com (2020). The truth about dog people. [Blog post] https://www.rover .com/blog/the-truth-about-dog-people-infographic/.

80 *Dan Gilbert of Harvard University claims that every psychologist*: Gilbert, D. (2006). *Stumbling on happiness.* New York, NY: Alfred A. Knopf.

81 *inevitably end in the death of the "playmate"*: Matsuzawa, T., Hockings, K. J., Humle, T., & Carvalho, S. (2012). Chimpanzee interactions with nonhuman species in an anthropogenic habitat. *Behaviour,* 149(3–4), 299–324.

81 *while visiting a biological preserve in Brazil*: Izar, P., Verderane, M. P., Visalberghi, E., Ottoni, E. B., Gomes De Oliveira, M., Shirley, J., & Fragaszy, D. (2006). Cross-genus adoption of a marmoset (Callithrix jacchus) by wild capuchin monkeys (Cebus libidinosus): Case report. *American Journal of Primatology,* 68(7), 692–700.

82 *two other examples of long-term, pet-like adoptions*: Carzon, P., Delfour, F., Dudzinski, K., Oremus, M., & Clua, É. (2019) Cross-genus adoptions in delphinids: One example with taxonomic discussion. *Ethology*, 125, 669–676. Mittal, D., Chakrabarti, S., Khambda, S. B., & Bump, J. K. (2020). Spots and manes: the curious case of foster care between two competing felids. *Ecosphere*, 11(2).

83 *breastfeeding of young animals by human females has been reported nearly worldwide*: Simoons, F. J., & Baldwin, J. A. (1982). Breast-feeding of animals by women: its socio-cultural context and geographic occurrence. *Anthropos*, 421–448.

83 *The psychologist Mike Tomasello*: Tomasello, M. (2009). *The cultural origins of human cognition.* Cambridge, MA: Harvard University Press.

84 *orgasm in human females*: Lloyd, E. A. (2005). *The case of the female orgasm: Bias in the science of evolution.* Cambridge, MA: Harvard University Press.

85 *a fundamental attribute of human nature*: Serpell, J. (1996). *In the company of animals: A study of human-animal relationships.* Cambridge, UK: Cambridge University Press. (p. 148).

85 *nearly 400 human universals*: Brown, D. E. (1991). *Human universals.* Philadelphia: Temple University Press. Pinker, S. (2002). *The blank slate: The modern denial of human nature.* New York, NY: Viking.

86 *nest parasitism, a reproductive strategy*: The idea that pets are a kind of nest parasitism was discussed by Serpell (1996) and by Archer, J. (1997). Why do people love their pets? *Evolution and Human Behavior*, 18(4), 237–259.

89 *how pet-keeping played out in sixty societies*: Gray, P. B., & Young, S. M. (2011). Human–pet dynamics in cross-cultural perspective. *Anthrozoös*, 24(1), 17–30.

89 *the pets of choice in Japanese homes*: Bulliet, R. W. (2005). *Hunters, herders, and hamburgers: The past and future of human-animal relationships.* New York, NY: Columbia University Press.

90 *"regretting the holocaust of pets that occurred at the outbreak of the war"*: Kean, H. (2017). *The great cat and dog massacre: The real story of world war two's unknown tragedy.* Chicago, IL: University of Chicago Press.

CHAPTER 4

93 *wagging tails, and kisses say it all*: Email from Ruby Benjamin (2009, August 10).

93 *fun out of owning one*: This statement was uttered by Bob Dylan during his XM satellite radio show, "Theme Time Radio Hour." The topic of the show was songs about dogs.

99 *Molecular biologists Heidi Parker and Elaine Ostrander*: Parker, H. G., & Ostrander, E. A. (2005). Canine genomics and genetics: Running with the pack. *PLoS Genetics*, 1(5), e58.

99 *Paleolithic cave paintings of animals go back 45,000 years*: Brumm, A., et al. (2021). Oldest caves are found in Sulawesi. *Science Advances.* https://advances.sciencemag.org/content/7/3/eabd4648 (Downloaded January 16, 2021).

100 *show up as early as 14,000 years ago*: Francis, R. C. (2015). *Domesticated: Evolution in a man-made world.* New York, NY: Norton.

100 *Domestication changes a species*: Clutton-Brock, J. (2017). Origins of the dog: Domestication and early history. In J. Serpell (Ed.), *The domestic dog: Its evolution, behaviour, and interactions with people* (pp. 7–20). Cambridge, UK: Cambridge University Press.

100 *New evidence suggests*: Sinding, M. H. S., Gopalakrishnan, S., Ramos-Madrigal, J., de Manuel, M., Pitulko, V. V., Kuderna, L., & Gilbert, M. T. P. (2020). Arctic-adapted dogs emerged at the Pleistocene–Holocene transition. *Science, 368*(6498), 1495–1499.

101 *In 1997, an article appeared in the journal* Science: Vila, C., Savolainen, P., Maldonado, J. E., Amorim, I. R., Rice, J. E., Honeycutt, R. L., et al. (1997). Multiple and ancient origins of the domestic dog. *Science, 276*, 1687–1689.

102 *The late biologist and dog-sled racer Ray Coppinger*: Coppinger, R., & Coppinger, L. (2016). *What is a dog?* Chicago, IL: University of Chicago Press.

102 *even hand-reared wolves:* Topal, J., et al. (in press). The dog as a model for understanding human social behavior. *Advances in the study of behavior.*

103 *energy content for mammal tissue, vegetables, and fruit*: The idea that the consumption of human feces played a role in canine evolution was first suggested to me in a talk by Brian Hare of Duke University. Butler, J. R., Brown, W. Y., & Du Toit, J. T. (2018). Anthropogenic food subsidy to a commensal carnivore: the value and supply of human faeces in the diet of free-ranging dogs. *Animals, 8*(5), 67.

103 *the anthropologist Jeremy Koster*: Koster, J. M. (2008). Hunting with dogs in Nicaragua: an optimal foraging approach. *Current Anthropology, 49*(5), 935–944.

104 *the most widely distributed mammal on earth*: Francis, R. C. (2015). *Domesticated: Evolution in a man-made world.* New York, NY: Norton.

105 *from an extraordinary genetic experiment with foxes*: Trut, L. N. (1999). Early canid domestication: The farm-fox experiment. *American Scientist, 87*, 160–169.

105 *remarkable experiment on Artic foxes*: For a fascinating book-length treatment of the Siberian fox farm studies, see Dugatkin, L. A., Trut, L., & Trut, L. N. (2017). *How to tame a fox (and build a dog): Visionary scientists and a Siberian tale of jump-started evolution.* University of Chicago Press.

106 *changes in the prefrontal cortex of the tame foxes*: Dugatkin, L. A. (2018). The silver fox domestication experiment. *Evolution: Education and Outreach, 11*(1), 1–5.

107 *compared the ability of wolves and dogs:* Hare, B., Brown, M., Williamson, C., & Tomasello, M. (2002). The domestication of social cognition in dogs. *Science, 298*(5598), 1634–1636.

107 *hot debate:* For an objective overview of this argument see Morrell, V. (2009). Going to the dogs. *Science, 325*, 1062–1065.

107 *They have found wolf pups*: Miklosi, A. (2007). *Dog behaviour, evolution, and cognition*. Oxford, UK: Oxford University Press.

108 *They can, however, quickly learn to use human points to find food*: Wynne, C. (2019). *Dog is Love: How and why your dog loves you*. New York, NY: Houghton Mifflin Harcourt.

108 *my favorite, BICF2G630798942*: vonHoldt, B. M., Shuldiner, E., Koch, I. J., Kartzinel, R. Y., Hogan, A., Brubaker, L., Wanser, S., Stahler, D., Wynne, C. D. L., Ostrander, E. A., Sinsheimer, J. S., & Udell, M. A. R. (2017). Structural variants in genes associated with human Williams-Beuren syndrome underlie stereotypical hypersociability in domestic dogs. *Science Advances*, 3, e1700398.

108 *In 2018, 180 papers were published on the topic*: Aria, M., Alterisio, A., Scandurra, A., Pinelli, C., & D'Aniello, B. (2020). The scholar's best friend: research trends in dog cognitive and behavioral studies. *Animal Cognition*, 1–13.

108 *Examples abound.*: Pilley, J. W., & Hinzmann, H. (2014). *Chaser: Unlocking the genius of the dog who knows 1000 words*. New York, NY: Simon and Schuster. Topál, J., Byrne, R. W., Miklósi, A., & Csányi, V. (2006). Reproducing human actions and action sequences: "Do as I Do!" in a dog. *Animal Cognition*, 9(4), 355–367. Fugazza, C., Sommese, A., Pogány, Á., & Miklósi, Á. (2020). Did we find a copycat? Do as I do in a domestic cat (*Felis catus*). *Animal Cognition*, 1–11. Berns, G. S., Brooks, A. M., Spivak, M., & Levy, K. (2017). Functional MRI in awake dogs predicts suitability for assistance work. *Scientific Reports*, 7, 43704. Johnston, A. M., Holden, P. C., & Santos, L. R. (2017). Exploring the evolutionary origins of overimitation: a comparison across domesticated and non-domesticated canids. *Developmental Science*, 20(4), e12460.

109 *intentionally developed by humans within the last 150 years*: Serpell, J. A., & Duffy, D. L. (2014). Dog breeds and their behavior. In *Domestic dog cognition and behavior* (pp. 31–57). Springer, Berlin, Heidelberg.

109 *turned to "citizen science"*: Stewart, L., MacLean, E. L., Ivy, D., Woods, V., Cohen, E., Rodriguez, K., & Miklósi, Á. (2015). Citizen science as a new tool in dog cognition research. *PloS one*, 10(9), e0135176.

109 *James Serpell and his colleagues at the Center for the Interaction of Animals and Society*: Hsu, Y., & Serpell, J. A. (2003). Development and validation of a questionnaire for measuring behavior and temperament traits in pet dogs. *Journal of the American Veterinary Medical Association*, 223(9), 1293–1300.

110 *trainability scores for 1,500 dogs of eleven breeds*: Serpell, J. A., & Hsu, Y. (2005). Effects of breed, sex, and neuter status on trainability in dogs. *Anthrozoös*, 18(3), 196–207.

110 *trace them to specific genes and differences in brain anatomy*: MacLean, E. L., Snyder-Mackler, N., VonHoldt, B. M., & Serpell, J. A. (2019). Highly heritable and functionally relevant breed differences in dog behaviour. *Proceedings of the Royal Society B*, 286(1912), 20190716.

110 *Four and a half million Americans are bitten*: Sacks, J. J., Kresnow, M., & Houston, B. (1996). Dog bites: How big a problem? *Injury Prevention*, 2(1), 52–54.

110 *two-thirds of these deaths were attributed to one breed—pit bulls.*: McCarthy, N. (2018, September 18). America's most dangerous dog breeds. *Forbes*. https://www.forbes.com/sites/niallmccarthy/2018/09/13/americas-most-dangerous-dog-breeds-infographic/?sh=1eb8fd5462f8 (Downloaded December 30, 2020).

111 *bite strangers, turn on their owners, and pick fights with other dogs*: Duffy, D. L., Hsu, Y., & Serpell, J. A. (2008). Breed differences in canine aggression. *Applied Animal Behaviour Science*, 114(3–4), 441–460.

111 *In her book, Pit Bull: The Battle Over an American Icon*: Dickey, B. (2016). *Pit bull: The battle over an American icon*. New York, NY: Alfred Knopf.

111 *a fifty-eight-year-old evangelical minister*: This account of the attack on Billingsly is based on a report in *Westworld*, Denver's independent newspaper. Maher, J. J. (2009, September 24). For two decades, pit bulls have been public enemy #1 in Denver. *Westworld*. https://www.westword.com/news/for-two-decades-pit-bulls-have-been-public-enemy-1-in-denver-but-maybe-its-time-for-a-recount-5105359 (Downloaded December 30, 2020).

112 *A 2015 study examined the ability of animal shelter workers*: Olson, K. R., Levy, J. K., Norby, B., Crandall, M. M., Broadhurst, J. E., Jacks, S., & Zimmerman, M. S. (2015). Inconsistent identification of pit bull-type dogs by shelter staff. *The Veterinary Journal*, 206(2), 197–202.

113 *The New Yorker writer Malcolm Gladwell argues*: Gladwell, M. (2006, January 30). Troublemakers: What pit bulls can teach about profiling. *The New Yorker*.

113 *Between 1979 and 1990, Rottweilers killed six people*: Sacks, J. J., Kresnow, M., & Houston, B. (1996). Dog bites: how big a problem? *Injury Prevention*, 2, 52–54.

114 *eight Americans were killed by Rottweilers and eight by German shepherds*: Data from DogsBite.com. https://www.dogsbite.org/dog-bite-statistics-fatalities-2014.php (Downloaded December 30, 2020).

114 *spread around the globe as fast as a Paris Hilton sex tape*: O'Rourke, M. (2007, July 13). The Croc Epidemic: How a heinous synthetic shoe conquered the world. *Slate*.

115 *article on the evolutionary psychology of human-animal relationships*: Herzog, H. (2002). Darwinism and the study of human-animal interactions. *Society and Animals*, 10(4), 361–367.

115 *numbers of puppy registrations for each breed for the previous three years*: "Breeds by Year Recognized," American Kennel Club, https://www.akc.org/press-center/articles-resources/facts-and-stats/breeds-year-recognized/.

116 *Their article was about names that people give their babies*: Hahn, M. W., & Bentley, A. R. (2003). Drift as a mechanism for cultural change: An

example from baby names. *Proceedings of the Royal Society B: Biological Sciences*, 270, 120–123.

118 *compared power law graphs to a hockey stick*: Gladwell, M. (2006, February 13). Million dollar Murray. *The New Yorker*, pp. 96–107.

118 *"they usually imitate each other"*: This quote is cited in Stewart-Williams, S. (2018). *The ape that understood the universe: How the mind and culture evolve.* Cambridge University Press.

120 *a classic case of fashion trickle-down from the rich to the wanna-be-rich*: Ritvo, H. (1987). *The animal estate: The English and other creatures in the Victorian age.* Cambridge, MA: Harvard University Press. Grier, K. C. (2006). *Pets in America: A history.* Chapel Hill, NC: The University of North Carolina Press.

121 *devastating critique in the* Atlantic Monthly: Derr, M. (1990). The politics of dogs. *Atlantic Monthly*, 265(3), 49. Derr, M. (2004). *Dog's best friend: Annals of the dog-human relationship.* Chicago, IL: University of Chicago Press.

121 *annual registrations seem to have stabilized at about 500,000 a year*: B. Bonnett (personal communication, October 6, 2020).

123 *caused by breeding to standards dictated by kennel clubs*: Asher, L., Diesel, G., Summers, J. F., McGreevy, P. D., & Collins, L. M. (2009). Inherited defects in pedigree dogs. Part 1: Disorders related to breed standards. *The Veterinary Journal*, 182(3), 402–411.

123 *trace their ancestry to thirty-one animals*: Chase, K., Sargan, D., Miller, K., Ostrander, E., & Lark, K. (2006). Understanding the genetics of autoimmune disease: Two loci that regulate late onset Addison's disease in Portuguese water dogs. *International Journal of Immunogenetics*, 33(3), 179–184.

CHAPTER 5

129 *"natural" part of womanhood*: Luke, B. (2007). *Brutal: Manhood and the exploitation of animals.* Urbana, IL: University of Illinois Press (p. 15).

129 *no longer economically necessary*: Washburn, S. L, & Lancaster, C. S. (1968). The evolution of hunting. In R. B. Lee & I. DeVore (Eds.), *Man the hunter.* Chicago, Il: Aldine Publishing (p. 299).

129 *interviewed veterinary school students*: Herzog, H. A., Vore, T. L., & New, J. C. (1989). Conversations with veterinary students: Attitudes, ethics, and animals. *Anthrozoös*, 2, 181–188.

132 *amassing every study*: For a summary of these findings see Herzog, H. A. (2006). Gender differences in human-animal interactions. *Anthrozoös*, 20(1), 7–21.

132 *roughly equal numbers of men and women own companion animals*: Applebaum, J. W., Peek, C. W., & Zsembik, B. A. (2020). Examining US pet ownership using the General Social Survey. *Social Science Journal*, 1–10.

132 *"My pet means more to me than any of my friends"*: These items are from the Lexington Attachment to Pets Scale. Johnson, T., Garrity, T., & Stallones, L.

(1992). Psychometric evaluation of the Lexington Attachment to Pets Scale (LAPS). *Anthrozoös*, 5(3), 160–175.

132 *And this finding also applies to children*: Endenburg, N., van Lith, H. A., & Kirpensteijn, J. (2014). Longitudinal study of Dutch children's attachment to companion animals. *Society & Animals*, 22(4), 390–414. Muldoon, J. C., Williams, J. M., & Currie, C. (2019). Differences in boys' and girls' attachment to pets in early-mid adolescence. *Journal of Applied Developmental Psychology*, 62, 50–58.

133 *how they played with their dogs*: Prato-Previde, E., Fallani, G., & Valsecchi, P. (2006). Gender differences in owners interacting with pet dogs: An observational study. *Ethology*, 112(1), 64–73.

133 *owners interacted with their dogs in a variety of situations*: Cimarelli, G., Turcsán, B., Bánlaki, Z., Range, F., & Virányi, Z. (2016). Dog owners' interaction styles: Their components and associations with reactions of pet dogs to a social threat. *Frontiers in Psychology*, 7, 1979.

133 *Gail Melson, a developmental psychologist*: Melson, G. F., & Fogel, A. (1996). Parental perceptions of their children's involvement with household pets. *Anthrozoös*, 9, 95–106.

133 *than men to cute creatures*: Sprengelmeyer, R., Perrett, D., Fagan, E., Cornwell, R., Lobmaier, J., Sprengelmeyer, A., et al. (2009). The cutest little baby face: A hormonal link to sensitivity to cuteness in infant faces. *Psychological Science*, 20(2), 149–154. Miller, S. C., Kennedy, C. C., DeVoe, D. C., Hickey, M., Nelson, T., & Kogan, L. (2009). An examination of changes in oxytocin levels in men and women before and after interaction with a bonded dog. *Anthrozoös*, 22(1), 31–42. Fridlund, A., & MacDonald, M. (1998). Approaches to Goldie: A field study of human approach responses to canine juvenescence. *Anthrozoös*, 11, 95–100.

134 *Stephen Kellert*: Kellert, S. R. (1996). *The value of life: Biological diversity and human society*. Washington, D.C.: Island Press.; Kellert, S. R., & Berry, J. K. (1987). Attitudes, knowledge, and behaviors toward wildlife as affected by gender. *Wildlife Society Bulletin*, 15(3), 363–371.

134 *three times more common in women than men*: Fredrikson, M., Annas, P., Fischer, H. Å., & Wik, G. (1996). Gender and age differences in the prevalence of specific fears and phobias. *Behaviour Research and Therapy*, 34(1), 33–39.

134 *Many studies have now reported similar sex differences*: Fleischman, D.S. (2021). Animal ethics and evolutionary psychology. In T. Shackelford (Ed.), *The SAGE Handbook of Evolutionary Psychology*. SAGE.

135 *felt about using dogs and chimpanzees in experiments*: Pifer, L., Shimizu, K., & Pifer, R. (1994). Public attitudes toward animal research: Some international comparisons. *Society and Animals*, 2(2), 95–113.

135 *Swedish researchers*: Hagelin, J., Carlsson, H. E., & Hau, J. (2003). An overview of surveys on how people view animal experimentation: Some factors that may influence the outcome. *Public Understanding of Science*, 12(1), 67.

135 *The National Opinion Research Center*: This data is from the 1993 General Social Survey.

136 *in the United States and Great Britain*: French, R. D. (1975). *Antivivisection and medical science in Victorian society*. Princeton, NJ: Princeton University Press.

136 *A study of dog rescuers*: Markovits, A. S., & Queen, R. (2009). Women and the world of dog rescue: A case study of the state of Michigan. *Society and Animals*, 17, 325–342.

137 *stick to their vegetarian or vegan diets*: Rosenfeld, D. L. (2020). Gender differences in vegetarian identity: How men and women construe meatless dieting. *Food Quality and Preference*, 81, 103859.

137 *gender differences in concern for animals*: Hart L. A., & Melese-d'Hospital, P. (1989) The gender shift in the veterinary profession and attitudes towards animals: A survey and overview. *Journal of Veterinary Medical Education*, 16, 27–30.

139 *that over 70% of battered women*: Ascione, F. (1998). Battered women's reports of their partners' and their children's cruelty to animals. *Journal of Emotional Abuse*, 1(1), 119–133. Flynn, C. P. (2000). Woman's best friend: Pet abuse and the role of companion animals in the lives of battered women. *Violence Against Women*, 6(2), 162–177. For a recent review, see Ascione, F. R., et al. (2007). Battered pets and domestic violence: Animal abuse reported by women experiencing intimate violence and by non-abused women. *Violence Against Women*, 13, 354–373.

139 *Data compiled by Pet-abuse.com*: Gerbasi, K. C. (2004). Gender and nonhuman animal cruelty convictions: Data from pet-abuse.com. *Society and Animals*, 12(4), 359–365.

139 *in 70% of the 15,000 cases*: These statistics were downloaded from pet-abuse.com in January 2009.

140 *about 5,000 cases of hoarding*: Arluke, A., Patronek, G., Lockwood, R., & Cardona, A. (2017). Animal hoarding. In *The Palgrave international handbook of animal abuse studies* (pp. 107–129). London: Palgrave Macmillan.

140 *Seventy-five to eighty-five percent of hoarders are women*: Dozier, M. E., Bratiotis, C., Broadnax, D., Le, J., & Ayers, C. R. (2019). A description of 17 animal hoarding case files from animal control and a humane society. *Psychiatry Research*, 272, 365–368.

140 *"The Great Bunny Rescue of 2006"*: Weise, E. (2006, August 8). Rabbit rescue ends some bad hare days. *USA Today*.

140 *public health implications of animal hoarding*: Hoarding of Animals Research Consortium (2002). Health implications of animal hoarding. *Health & Social Work*, 27(2), 125–132.

141 *explanations of why people hoard animals*: Patronek, G., Loar, L., & Nathanson, J. N. (2006). *Animal hoarding: Structuring interdisciplinary responses to help people, animals, and communities at risk*. Boston: Hoarding of Animals Research Consortium.

141 *caused by toxoplasmosis infection*: Sklott, R. (2007, December 9). "Cat Lady" conundrum. *New York Times*.

141 *dog hoarders had lower rates of exposure to tox*: Cunha, G. R. D., Pellizzaro, M., Martins, C. M., Rocha, S. M., Yamakawa, A. C., Silva, E. C. D., & Biondo, A. W. (2020). Spatial serosurvey of anti-*Toxoplasma gondii* antibodies in individuals with animal hoarding disorder and their dogs in Southern Brazil. *PloS One*, 15(5), e0233305.

141 *The recidivism rate among hoarders*: Miller, S. (2008, January/February). Objects of their affection: The hidden world of hoarders. *Best Friends Magazine*, 20–22; 57–61. Non-animal hoarding has been linked to damage to the pre-frontal cortex of the brain; see Anderson, S. W., Damasio, H., & Damasio, A. R. (2005). A neural basis for collecting behaviour in humans. *Brain*, 128(1), 201–212. No one has tested the possibility that animal hoarding can be caused by brain damage.

142 *They contained between 387 and 697 cats*": Polak, K. C., Levy, J. K., Crawford, P. C., Leutenegger, C. M., & Moriello, K. A. (2014). Infectious diseases in large-scale cat hoarding investigations. *The Veterinary Journal*, 201(2), 189–195.

144 *A study involving over 300,000 people in 67 countries*: Atari, M., Lai, M. H., & Dehghani, M. (2020). Sex differences in moral judgements across 67 countries. *Proceedings of the Royal Society B*, 287(1937), 20201201.

144 *Some animal activists*: Gaarder, E. (2011). Where the boys aren't: The predominance of women in animal rights activism. *Feminist Formations*, 54–76.

144 *chimpanzees, males hunt more than females.*: Gilby, I. C., Machanda, Z. P., O'Malley, R. C., Murray, C. M., Lonsdorf, E. V., Walker, K., & Pusey, A. E. (2017). Predation by female chimpanzees: toward an understanding of sex differences in meat acquisition in the last common ancestor of Pan and Homo. *Journal of Human Evolution*, 110, 82–94.

144 *hunting is defined as a male activity in every human culture*: Gurven, M., Hill, K., Hames, R., Kameda, T., McDermott, R., Lupo, K., & Hill, K. (2009). Why do men hunt? A reevaluation of "man the hunter" and the sexual division of labor. *Current Anthropology*, 50(1), 51–74.

144 *nearly every human culture*: The following examples of women hunters in tribal society are from Estioko-Griffin, A., & Griffin, P. B. (1981). Woman the hunter: The Agta. In F. Dahlbrg (Ed.), *Woman the gatherer* (pp. 121–140). New Haven, CT: Yale University Press. Bailey, R. C., & Aunger, R. (1989). Net hunters vs. archers: Variation in women's subsistence strategies in the Ituri forest. *Human Ecology*, 17(3), 273–297. Romanoff, S. (1983). Women as hunters among the Matses of the Peruvian amazon. *Human Ecology*, 11(3), 339–343. Goodman, M. J., Griffin, P. B., Estioko-Griffin, A. A., & Grove, J. S. (1985). The compatibility of hunting and mothering among the Agta hunter-gatherers of the Philippines. *Sex Roles*, 12(11), 1199–1209.

145 *the result of socialization*: Borgi, M., & Cirulli, F. (2015). Attitudes toward animals among kindergarten children: species preferences. *Anthrozoös*, 28(1), 45–59. McClure, E. B. (2000). A meta-analytic review of sex differences in facial expression processing and their development in infants, children, and adolescents. *Psychological Bulletin*, 126(3), 424. Quinn, P. C., & Liben, L. S. (2008). A sex difference in mental rotation in young infants. *Psychological Science*, 19(11), 1067–1070. Moore, D. S., & Johnson, S. P. (2008). Mental rotation in infants. *Psychological Science*, 19(11), 1063–1066. Williams, C. L., & Pleil, K. E. (2008). Toy story: Why do monkey and human males prefer trucks? Comment on "Sex differences in rhesus monkey toy preferences parallel those of children" by Hassett, Siebert, and Wallen. *Hormones and Behavior*, 54, 355–358.

145 *make men more generous*: Domes, G., Heinrichs, M., Michel, A., Berger, C., & Herpertz, S. C. (2007). Oxytocin improves "Mind-reading" in humans. *Biological Psychiatry*, 61(6), 731–733. Zak, P. J., Stanton, A. A., & Ahmadi, S. (2007). Oxytocin increases generosity in humans. *PLoS ONE*, 2(11), e1128.

145 *glue that cements the human-animal bond*: Odendaal, J. S. (2000). Animal-assisted therapy—magic or medicine? *Journal of Psychosomatic Research*, 49(4), 275–280. Miller, S. C., Kennedy, C., DeVoe, D., Hickey, M., Nelson, T., & Kogan, L. (2009). An examination of changes in oxytocin levels in men and women before and after interaction with a bonded dog. *Anthrozoös*, 22(1), 31–42. Nagasawa, M., Kikusui, T., Onaka, T., & Ohta, M. (2009). Dog's gaze at its owner increases owner's urinary oxytocin during social interaction. *Hormones and Behavior*, 55(3), 434–441. Powell, L., Edwards, K. M., Bauman, A., Guastella, A. J., Drayton, B., Stamatakis, E., & McGreevy, P. (2019). Canine endogenous oxytocin responses to dog-walking and affiliative human–dog interactions. *Animals*, 9(2), 51.

146 *pets have reported mixed results*: Powell, L., Guastella, A. J., McGreevy, P., Bauman, A., Edwards, K. M., & Stamatakis, E. (2019). The physiological function of oxytocin in humans and its acute response to human-dog interactions: A review of the literature. *Journal of Veterinary Behavior*, 30, 25–32.

147 *Height is a good example.*: The height example is taken from Pinker, S. (2003). *The blank slate: The modern denial of human nature*. New York, NY: Penguin.

CHAPTER 6

149 *just secondhand cowards*: Dundes, A. (Ed.) (1994). *The cockfight: A casebook*. Madison, WI: University of Wisconsin Press. (p. 71).

149 *humane, sport there is*: Fitz-Barnard, L. (1921). *Fighting sports*. London: Oldam Press (p. 12).

154 *who tried to make sense of rooster fighting*: For a compendium of historical and anthropological writings on the significance of cockfighting in different

cultures see Dundes, A. (Ed.) (1994). *The cockfight: A casebook*. Madison, WI: University of Wisconsin Press.

154 *"homoerotic male battle with masturbatory nuances"*: Dundes, A. (1994). The gallus as phallus. In Dundes, A. (Ed.), *The cockfight: A casebook*. (pp. 241–281). Madison, WI: University of Wisconsin Press.

155 *oldest and most widespread of traditional sports*: For a history of cockfighting see Smith, P., & Daniel, C. (2000). *The chicken book*. Athens, GA: University of Georgia Press.

155 *date back to 517 BCE in China*: Lawler, A. (2014). *Why did the chicken cross the world?* New York, NY: Atria Press.

155 *in Spain, France, the British Isles, and Scandinavia*: Walker, S. J., & Meijer, H. J. (2020). More than food; evidence for different breeds and cockfighting in gallus bones from Medieval and Post-Medieval Norway. *Quaternary International*.

157 *the basic set of guidelines is referred to as Wortham's Rules*: Wortham's Rules apply to the gaff fights that were preferred by Appalachian cockfighters. Hispanic cockfighters typically use a different set of rules.

157 *"a Platonic concept of hate"*: Geertz, C. (1994). Deep play: Notes on the Balinese cockfight. In A. Dundes (Ed.), *The cockfight: A casebook*. (p. 103). Madison, WI: University of Wisconsin Press.

161 *the sociologist Clifton Bryant estimated*: Bryant, C. (1991). Deviant leisure and clandestine lifestyle: Cockfighting as a socially disvalued sport. *World Leisure and Recreation*, 33(2), 17–21.

163 *Captain L. Fitz-Barnard claimed*: Fitz-Barnard, L. (1921). *Fighting Sports*. London: Oldam Press (p. 12).

165 *the Humane Society of the United States has linked cockfighting with*: Website http://www.hsus.org/acf/fighting/cockfight/cockfighting_and_related_crimes.html (Downloaded March 12, 2009).

166 *"That's what's kept me interested in it"*: The quote is by Bobby Keener of Greensboro, North Carolina. It is from the dvd, *Cockfighters*, an eight hour series of interviews with cockfighters produced by Olena Media http://www.olenamedia.com/.

166 *cockfighting magazine* Grit and Steel: McCaghy, C. H., & Neal, A. G. (1974). The fraternity of cockfighters: Ethical embellishments of an illegal sport. *Journal of Popular Culture*, (8), 557–569.

168 *She is the founder of United Poultry Concerns*: Davis, K. (2009). *Prisoned chickens, poisoned eggs: An inside look at the modern poultry industry*. Summertown, TN: Book Publishing Company.

170 *their trip to the processing plant*: Eisnitz, G. A. (2007). *Slaughterhouse: The shocking story of greed, neglect, and inhumane treatment inside the U.S. meat industry*. Amherst, NY: Prometheus Books.

170 *The Humane Society of the United States*: Humane Society of the United States (2009). *An HSUS Report: Welfare Issues with Conventional Manual Catching of Broiler Chickens and Turkeys*.

171 *from 140 birds per minute to 175 birds*: Kindly, K. Mellnik, T., & Hernandez, A. R. (2021, January 3). The Trump Administration Approved Line Speeds at Chicken Plants. *Washington Post*. Wabeck, Bell, D. D., & Weaver, W. D. (2001). *Commercial chicken meat and egg production*. New York, NY: Springer.

171 *and, finally, bagger*: U.S. Department of Labor. Poultry Processing Industry eTool report. https://www.osha.gov/SLTC/etools/poultry/evisceration.html (Downloaded November 3, 2020).

173 *the $70 billion poultry industry*: Lawler, A. (2014). *Why did the chicken cross the world?* New York, NY: Atria.

173 *The eighteenth century movement against bloodsports*: See Thomas, K. (1983). *Man and the natural world: a history of the modern sensibility*. New York, NY: Pantheon.

174 *essay that appeared in the* Atlanta Journal-Constitution: Rudy, K. (2007, September 6). Dog-fighting and Michael Vick. *Atlanta Journal Constitution*.

174 *You see the same thing in thoroughbred racing.*: McMurray, J. (2008, June 14). AP finds 5K horse deaths since '03. *Washingtonpost.com*.

174 *About 1,000 horses a year die*: Battuello, P. Website: Horseracing Wrongs. Website https://horseracingwrongs.org/ (Downloaded November 3, 2020).

174 *support a ban on thoroughbred horseracing*: Paulik Report staff (2019). Reuters/Ipsos public opinion poll shows challenges ahead for horse racing. *Paulik Report*. https://www.paulickreport.com/news/the-biz/reuters-ipsos-public-opinion-poll-shows-challenges-ahead-for-horse-racing/.

CHAPTER 7

177 *the demand for bacon*: Singer, P. (1993). *Practical ethics* (second edition). New York: Cambridge University Press.

178 *"less about your cholesterol than your sanity"*: Bruni, F. (2009, May 20). Beef and Décor, Aged to Perfection. *New York Times*.

179 *Chimpanzees, our closest living relatives, love the taste of meat.*: For more information on chimpanzee carnivory see Stanford, C. B. (1999). *The hunting ape: Meat eating and the origins of human behavior*. Princeton, NJ: Princeton University Press.; Stanford, C. (2002). *Significant others: The ape-human continuum and the quest for human nature*. New York: Basic Books. Boesch, C. (1994). Hunting strategies of Gombe and Taï chimpanzees. In R. W. Wrangham, W. C. McGrew, F. B. M. de Waal, & P. G. Heltne (Eds.), *Chimpanzee cultures* (pp. 76–92). Cambridge, MA: Harvard University Press. Foley, R. (2001). The evolutionary consequences of increased carnivory. In C. B. Stanford & H. T. Bunn (Eds.), *Meat-eating and human evolution* (pp. 305–331). New York: Oxford University Press. The exchange of meat for sex is described in Gomes, C. M., & Boesch, C. (2009). Wild chimpanzees exchange meat for sex on a long-term basis. *PLoS ONE*, 4, e5116. Fedurek, P., Tkaczynski, P., Asiimwe, C., Hobaiter,

C., Samuni, L., Lowe, A. E., & Crockford, C. (2019). Maternal cannibalism in two populations of wild chimpanzees. *Primates*,1–7.

180 *mothers occasionally eat their own infants*: Fedurek, P., Tkaczynski, P., Asiimwe, C., Hobaiter, C., Samuni, L., Lowe, A. E., & Crockford, C. (2019). Maternal cannibalism in two populations of wild chimpanzees. *Primates*, 1–7.

181 *"greedily devouring livid writhing flesh"*: Stanford, C. B. (1999). *The hunting ape: Meat eating and the origins of human behavior*. Princeton, NJ: Princeton University Press (p. 107).

181 *the Nunamuit people of northern Alaska*: Cordain, L., Eaton, S., Brand Miller, J., Mann, N., & Hill, K. (2002). Original communications—the paradoxical nature of hunter-gatherer diets: Meat-based, yet non-atherogenic. *European Journal of Clinical Nutrition*, 56(1), 42–52. Gadsby, P. (2004, October 1). The Inuit paradox. *Discover Magazine*.

181 *a diet composed of less than 15% animal products*: Cordain, L., Eaton, S. B., Sebastian, A., Mann, N., Lindeberg, S., Watkins, B. A., et al. (2005). Origins and evolution of the Western diet: Health implications for the 21st century. *American Journal of Clinical Nutrition*, 81(2), 341–354.

183 *Psychologists call this the "meat paradox"*: Bastian, B., & Loughnan, S. (2017). Resolving the Meat-Paradox: A motivational account of morally troublesome behavior and its maintenance. *Personality and Social Psychology Review*, 21(3), 278–299.

183 *these are referred to as the "4Ns"*: Joy, M. (2010). *Why we love dogs, eat pigs, and wear cows: An introduction to carnism*. Conari Press. Piazza, J., Ruby, M. B., Loughnan, S., Luong, M., Kulik, J., Watkins, H. M., & Seigerman, M. (2015). Rationalizing meat consumption. The 4Ns. *Appetite*, 91, 114–128.

184 *humans are unique among animals in spicing the food we eat*: Rozin, P., & Schiller, D. (1980). The nature and acquisition of a preference for chili pepper by humans. *Motivation and Emotion*, 4(1), 77–101. Billing, J., & Sherman, P. W. (1998). Antimicrobial functions of spices: why some like it hot. *The Quarterly Review of Biology*, 73(1), 3–49.

184 *especially for pregnant women*: Fessler, D. M. T., Bayley, T. M., Dye, L., Brown, J. K., Flaxman, S. M., Leeners, B., et al. (2002). Reproductive immunosuppression and diet. *Current Anthropology*, 43(1), 19–61. Flaxman, S. M., & Sherman, P. W. (2000). Morning sickness: A mechanism for protecting mother and embryo. *Quarterly Review of Biology*, 75(2), 113–148.

187 *six times more common than fruit aversions*: Midkiff, E. E., & Bernstein, I. L. (1985). Targets of learned food aversions in humans. *Physiology & Behavior*, 34(5), 839–841.

187 *food taboos across human societies*: Fessler, D., & Navarrete, C. (2003). Meat is good to taboo: Dietary proscriptions as a product of the interaction of psychological mechanisms and social processes. *Journal of Cognition and Culture*, 3(1), 1–40.

188 *a taboo against eating buffalo meat*: McDonaugh, C. (1997). Breaking the rules: Changes in food acceptability among the Tharu of Nepal. In H. Macbeth (Ed.), *Food Preferences and Taste: Continuity and Change*. Oxford, UK: Berghahn Books.

188 *humans have been eating dogs for thousands of years*: McHugh, S. (2004). *Dog*. London, GK: Reaktion Books. Serpell, J. (1995). The hair of the dog. In J. Serpell (Ed.), *The domestic dog: Its evolution, behavior and interactions with people*. (pp. 257–262). Cambridge, UK: Cambridge University Press. Pringle, H. (2011, May 6). Earliest American dogs may have been dinner. *Science*. https://www.sciencemag.org/news/2011/05/earliest-american-dogs-may-have-been-dinner.

189 *The anthropologist Frederick Simmons*: Simmons, F. J. (1996). Dogflesh eating by humans in sub-Saharan Africa. *Ecology of Food and Nutrition*, 34(4), 251–291.

189 *the Asian trade in dog products*: Czajkowski, C. (2014). Dog meat trade in South Korea: A report on the current state of the trade and efforts to eliminate it. *Animal Law*, 21, 29–64. Podberscek, A. (2009). Good to pet and eat: The keeping and consuming of dogs and cats in South Korea. *Journal of Social Issues*, 65, 615–632. Walraven, B. (2002). Bardot soup and Confucians' meat: Food and Korean identity in global context. In K. Cwiertka & B. Walraven (Eds.), *Asian food: The global and the local* (pp. 95–115). Honolulu: University of Hawaii Press.

190 *dogmeat is also dimming in China*: Thompson, B. (2020, May 29). China signals complete ban on the eating of dog meat. *Daily Mail*.

190 *In classical Hinduism*: Nelson, L. (2006). Cows, elephants, dogs, and other lesser embodiments of Atman: Reflections of Hindu attitudes toward nonhuman animals. In Waldau, P., & Patton, K. *A communion of subjects: Animals in religion, science, and ethics* (pp. 179–193). New York, NY: Columbia University Press.

190 *Islamic law also regard dogs as unclean*: Foltz, R. (2006). "This she-camel of God is a sign to you": Dimensions of animals in Islamic tradition and Muslim culture. In P. Waldau & K. Patton (Eds.), *A communion of subjects: Animals in religion, science, and ethics* (pp. 149–150). New York: Columbia University Press.

191 *Oglala Indians of the Pine Ridge Reservation in South Dakota*: Powers, W., & Powers, M. (1986). Putting on the dog. *Natural History*, 2, 6–16.

191 *hunters atone for killing animals are nearly universal*: Serpell, J. (1996). *In the company of Animals: A study of human-animal relationships* (second edition), Cambridge, UK: Cambridge University Press.

191 *animal liberation philosopher Peter Singer would probably not object to*: Singer, P., & Mason, J. (2006). *The ethics of what we eat: Why our food choices matter*. Emmaus, PA: Rodale Press.

192 *Sandy and I distributed questionnaires*: Herzog, H., & McGee, S. (1983). Psychological aspects of slaughter: Reactions of college students of killing

and butchering cattle and hogs. *International Journal for the Study of Animal Problems*, 4(2), 124–132.

193 *He wrote, "Humans must eat, excrete, and have sex"*: Rozin, P., Haidt, J., & McCauley, C. R. (2000). Disgust. In M. Lewis & J.M. Haviland-Jones (Eds.), *Handbook of emotions* (second edition). New York, NY: Guilford Press. (p. 642).

193 *For instance, researchers*: Kubberød, E., Ueland, Ø., Dingstad, G. I., Risvik, E., & Henjesand, I. J. (2008). The effect of animality in the consumption experience: A potential for disgust. *Journal of Food Products Marketing*, 14(3), 103–124. Kubberød, E., Ueland, Ø., Rødbotten, M., Westad, F., & Risvik, E. (2002). Gender specific preferences and attitudes towards meat. *Food Quality and Preference*, 13(5), 285–294.

194 *immoral is called "moralization"*: Rozin, P., Markwith, M., & Stoess, C. (1997). Moralization and becoming a vegetarian: The transformation of preferences into values and the recruitment of disgust. *Psychological Science*, 8, 67–73.

195 *important factor in their purchasing decisions.*: Spain, C. V., Freund, D., Mohan-Gibbons, H., Meadow, R. G., & Beacham, L. (2018). Are they buying it? United States consumers' changing attitudes toward more humanely raised meat, eggs, and dairy. *Animals*, 8(8), 128.

195 *the Vegetarian Resource Group*: The results of the Vegetarian Resource Group surveys can be found at https://www.vrg.org/nutshell/Polls/2019 _adults_veg.htm.

197 *A multisite study of half a million people.*: Sinha, R., Cross, A. J., Graubard, B. I., Leitzmann, M. F., & Schatzkin, A. (2009). Meat intake and mortality: A prospective study of over half a million people. *Archives of Internal Medicine*, 169, 562–571. Schwingshackl, L., Schwedhelm, C., Hoffmann, G., Lampousi, A. M., Knüppel, S., Iqbal, K., & Boeing, H. (2017). Food groups and risk of all-cause mortality: a systematic review and meta-analysis of prospective studies. *The American Journal of Clinical Nutrition*, 105(6), 1462–1473.

197 *Here's how Newkirk explained PETA's logic to me*: Ingrid Newkirk (email) June 24, 2009.

199 *seafood within the previous twenty-four hours*: Juan, W., Yamini, S., & Britten, P. (2015). Food intake patterns of self-identified vegetarians among the U.S. population, 2007–2010. *Procedia Food Science*, 4, 86–93.

201 *In the United States, vegetarians*: Maurer, D. (2010). *Vegetarianism: Movement or moment: Promoting a lifestyle for cult change*. Philadelphia: PA: Temple University Press. Forestell, C. A., & Nezlek, J. B. (2018). Vegetarianism, depression, and the five factor model of personality. *Ecology of Food and Nutrition*, 57(3), 246–259. Nezlek, J. B., & Forestell, C. A. (2019). Where the rubber meats the road: Relationships between vegetarianism and socio-political attitudes and voting behavior. *Ecology of Food and Nutrition*, 58(6), 548–559. Millum, J. (2018, June 20). Who are the vegetarians.

(webpage) *Faunalytics* https://faunalytics.org/who-are-the-vegetarians/. Heiss, S., & Hormes, J. M. (2018). Ethical concerns regarding animal use mediate the relationship between variety of pets owned in childhood and vegetarianism in adulthood. *Appetite*, 123, 43–48.

202 *some studies have linked vegetarianism, particularly in women, to eating disorders*: Sergentanis, T. N., Chelmi, M. E., Liampas, A., Yfanti, C. M., Panagouli, E., Vlachopapadopoulou, E., & Tsitsika, A. (2021). Vegetarian diets and eating disorders in adolescents and young adults: A systematic review. *Children*, 8(1), 12. Dorard, G., & Mathieu, S. (2020). Vegetarian and omnivorous diets: A cross-sectional study of motivation, eating disorders, and body shape perception. *Appetite*, 156, 104972.

202 *which compared rates of depression in meat-eaters and meat-avoiders*: For a summary of these papers, see Herzog, H. (2019). The sad truth about pet ownership and depression. (blog post) *Animals and Us.* https://www.psychologytoday.com/us/blog/animals-and-us/201912/the-sad-truth-about-pet-ownership-and-depression (Downloaded November 25, 2020).

203 *systematic review to help make sense of inconsistent results*: Dobersek, U., Wy, G., Adkins, J., Altmeyer, S., Krout, K., Lavie, C. J., & Archer, E. (2020). Meat and mental health: a systematic review of meat abstention and depression, anxiety, and related phenomena. *Critical Reviews in Food Science and Nutrition*, 1–14. Iguacel, I., Huybrechts, I., Moreno, L. A., & Michels, N. (2020). Vegetarianism and veganism compared with mental health and cognitive outcomes: a systematic review and meta-analysis. *Nutrition Reviews.*

205 *Twenty-five percent of vegans*: MacInnis, C. C., & Hodson, G. (2017). It ain't easy eating greens: Evidence of bias toward vegetarians and vegans from both source and target. *Group Processes & Intergroup Relations*, 20(6), 721–744.

206 *survey of over 11,000 American adults*: Humane Research Council (2014). Study of Current and Former Vegetarians: Initial Findings. https://faunalytics.org/wp-content/uploads/2015/06/Faunalytics_Current-Former-Vegetarians_Full-Report.pdf.

206 *Masson extolled the health benefits*: Masson, J. (2010). *The face on your plate: The truth about food.* New York: W. W. Norton & Company (p. 168).

207 *"become a constant ordeal"*: Steiner, G. (2009, November 22). Animal, vegetable, miserable. *New York Times.*

208 *each time I ordered a hamburger*: Haidt, J. (2006). *The happiness hypothesis.* New York, NY: Basic Books (p. 165).

208 *"the no-man's land in the battle between mind and body"*: This line is taken from the 1999 movie *The Devil's Advocate.*

209 *the cognitive chasm between humans and chimpanzees*: Marc Hauser is quoted as saying this in Balter, M. (2008). How human intelligence evolved—Is it science or "paleofancy"? *Science*, 319, 1028.

CHAPTER 8

211 *Both alternatives are absurd.*: Cohen, C. (1987). The case for the use of animals in biomedical research. *New England Journal of Medicine,* 315, 865–870 (p. 867).

211 *the human body writ small*: Goodman, A. (2006). *Intuition: A novel.* New York, NY: Dial Press.

215 *angelic little Fan-tail Pointer at ten days old*: Danta, C. (2018). "The Highest Civilisation among Ants": Stevenson and the Fable. In *Animal fables after Darwin: Literature, speciesism, and metaphor* (pp. 61–95). Cambridge: Cambridge University Press. doi:10.10^{17}/₉₇81108552394.004.

216 *"for mere damnable and detestable curiosity"*: Browne, J. (2002). *Charles Darwin: The power of place.* Princeton, NJ: Princeton University Press (p. 421).

216 *he amended the sentence*: Both versions are cited in Burghardt, G. M., & Herzog, H. A. (1980). Commentary: Beyond conspecifics: Is Brer Rabbit our brother? *Bioscience,* 30, 763–768.

216 *Claude Bernard who wrote*: Rudacille, D. (2000). *The scalpel and the butterfly: The war between animal research and animal protection.* New York, NY: Farrar, Straus, and Giroux. (p. 36).

217 *all of them said yes when it came to pain*: Herzog, H. (1991). Animal consciousness and human conscience. *Contemporary Psychology,* 36, 7–8.

217 *a more systematic survey*: Knight, S., Vrij, A, Bard, K, & Brandon, D. (2009). Science versus human welfare: Understanding attitudes toward animal use. *Journal of Social Issues,* 65, 463–483.

223 *the biological equivalent of a Swiss Army knife*: Schipani, S. (2019, February 27). The history of the lab rat is full of scientific triumphs and ethical quandaries. *Smithsonian Magazine.* https://www.smithsonianmag.com/science-nature/history-lab-rat-scientific-triumphs-ethical-quandaries-180971533/ (Downloaded January 24, 2020).

223 *According to a Zogby poll*: The Zogby poll was conducted in February 2009 and was commissioned by the Foundation for Biomedical Research.

223 *transformation of the mouse from pest to pet to model organism*: Rader, K. A. (2004). *Making mice: Standardizing animals for American biomedical research, 1900–1955.* Princeton, NJ: Princeton University Press.

223 *eugenics advocate Clarence Little*: Critser, G. (2007, December). Of Mice and Men: How a twenty-gram rodent conquered the world of science. *Harper's Magazine,* 65–76.

224 *engineer a strain of mice with humanized ACE2*: Cohen, J. (2020, April 13). From mice to monkeys, animals studied for coronavirus answers. *Science,* 221–222.

225 *99.5% of mouse genes have a known human counterpart*: Paigen, K. (2003). One hundred years of mouse genetics: An intellectual history. II. The molecular revolution (1981–2002). *Genetics,* 163(4), 1227–1235.

225 *According to Rick Woychik*: This quote is taken from a Jackson Laboratory promotional video (http://www.jax.org/).

226 *The ethicist Carl Cohen*: Cohen, C. (1987). The case for the use of animals in biomedical research. *New England Journal of Medicine*, 315, 865–870 (p. 868).

227 *the problem of replication of research results*: Schipani, S. (2019, February 27). The history of the lab rat is full of scientific triumphs and ethical quandaries. *Smithsonian Magazine*. https://www.smithsonianmag.com/science -nature/history-lab-rat-scientific-triumphs-ethical-quandaries-180971533 / (Downloaded December 2, 2020).

227 *Researchers in Portland, Edmonton, and Albany ran eight strains of mice*: Crabbe, J. C., Wahlsten, D., & Dudek, B. C. (1999). Genetics of mouse behavior: Interactions with laboratory environment. *Science*, 284, 1670–1672.

228 *Writing in the journal* Immunity: Davis, M. M. (2008). A prescription for human immunology. *Immunity*, 29(6), 835–838.

228 *drug that worked in four mouse studies made people with ALS sicker*: Schnabel, J. (2008). Neuroscience: Standard model. *Nature*, 454(7205), 682–685.

228 *Only one in ten cancer treatments*: Dennis, C. (2006). Off by a whisker. *Nature*, 441, 739–741.

228 *The list of reasons experiments on mice fail*: Van der Worp, H. B., Howells, D. W., Sena, E. S., Porritt, M. J., Rewell, S., O'Collins, V., & Macleod, M. R. (2010). Can animal models of disease reliably inform human studies? *PLoS Medicine*, 7(3), e1000245.

232 *over 99% of the animals used in research*: Carbone, L. (2021). Estimating mouse and rats use in American laboratories by extrapolation form Animal Welfare Act-regulated species. *Scientic Reports*.

233 *completely exempt from the regulations*: Note that birds, rats, and mice intended for research in institutions which receive funding from the National Institutes of Health are covered under the Public Health Service *Guide for the Care and Use of Laboratory Animals*, a separate set of government animal care regulations. However, mice in an estimated 800 research facilities are not covered under any federal regulations.

233 *Dr. Larry Carbone used the Freedom of Information Act*: Carbone, L. (2021). Estimating mouse and rat use in American laboratories by extrapolation from Animal Welfare Act-regulated species. *Scientific Reports*.

235 *made their decisions by flipping a coin*: Plous, S., & Herzog, H. (2001). Reliability of protocol reviews for animal research. *Science*, 293(5530), 608–609.

236 *In his novel* Zen and the Art of Motorcycle Maintenance: Pirsig, R. (1974). *Zen and the art of motorcycle maintenance*. New York, NY: William Morrow (p. 185).

236 *quality of research abstracts submitted for presentation at anthrozoology conferences*: Herzog, H. A., Podberscek, A. L., & Docherty, A. (2005). The reliability of peer review in anthrozoology. *Anthrozoös*, 18(2), 175–182.

238 *"imitate actions and vocalizations"*: Wise, S. M. (2002). *Drawing the line: Science and the case for animal rights.* Cambridge, MA: Perseus Books (p. 157).

239 *capable of empathy*: Langford, D. J., Crager, S. E., Shehzad, Z., Smith, S. B., Sotocinal, S. G., Levenstadt, J. S., et al. (2006). Social modulation of pain as evidence for empathy in mice. *Science, 312*(5782), 1967–1970.

239 *If I calculated correctly, the research involved over 800 mice*: It is difficult to actually figure out how many total animals were used in the experiments from research reports that appeared in *Science.* In addition to the published article, there were thirty additional pages of supplementary materials published online.

241 *voice for animal protection*: Bekoff made this statement in a speech to the Farm Sanctuary Hoe Down (www.farmsanctuary.org) in Orland, California, on May 16, 2009.

242 *the Nazi medical experiments*: For discussions of the ethics of using the Nazi medical data see Moe, K. (1984, December). Should the Nazi research data be cited? *Hastings Center Report,* 5–7. and Cohen, B. (1990). The ethics of using medical data from Nazi experiments. *Journal of Halacha and Contemporary Society, 19,* 103–126.

242 *experiments on animals are also ill-gotten gains*: See, for example, Regan, T. (1993). Ill gotten gains. In P. Cavalieri & P. Singer (Eds.), *The great ape project* (pp 194–205). New York, NY: St. Martin's Press.

CHAPTER 9

247 *and how can this be determined?*: Nozick, R. (1974). *Anarchy, state and utopia.* New York, NY: Basic Books.

247 *hypocrisy we only compound our own.*: Haidt, J. (2006). *The happiness hypothesis: Finding modern truth in ancient wisdom.* New York, NY: Basic Books.

248 *"Killing animals so you can toss their bodies"*: Byrne's quote appears in Murphy, K. (2009, June 13). Seattle's Pike Place fishmongers under fire. *Los Angeles Times.*

248 The Powerful Bond Between People and Pets: Anderson, P. E. (2008). *The powerful bond between people and pets: Our boundless connections to companion animals.* Westport, CT: Praeger (p. 214).

248 *70% of animal activists*: Plous, S. (1991). An attitude survey of animal rights activists. *Psychological Science, 2*(3), 194–196.

249 *an article by an ethicist named Mylan Engel twenty years ago*: Engel, M. (2000). The immorality of eating meat. In L. P. Pojman (Ed.), *The moral life: An introductory reader in ethics and literature* (pp. 856–889). New York: Oxford University Press.

250 *"non-attitudes" or "vacuous attitudes"*: Sturgis, P., & Smith, P. (2010). Fictitious issues revisited: Political interest, knowledge and the generation of nonattitudes. *Political Studies, 58*(1), 66–84. Herzog, H., Rowan, A., &

Kossow, D. (2001). Social attitudes and animals. In A. N. Rowan & D. J. Salem (Eds.), *The state of the animals* (pp. 55–69). Washington, D.C.: Humane Society Press.

250 *the majority of people do not get in a much of a twit*: The data in this paragraph are from a report by the Humane Research Council (2004, March). *Understanding the public image of the U.S. animal protection movement*.

251 *"psychic numbing"*: Resnick, B. (2017) A psychologist explains the limits of human compassion. *Vox*. https://www.vox.com/explainers/2017/7/19 /15925506/psychic-numbing-paul-slovic-apathy (Downloaded December, 7, 2010). Slovic, P. (2007). If I look at the mass I will never act: Psychic numbing and genocide. *Judgment and Decision Making*, 2(2), 79–95.

252 *most common thread was moral shock*: Jasper, J. M., & Poulsen, J. D. (1995). Recruiting strangers and friends: Moral shocks and social networks in animal rights and anti-nuclear protests. *Social Problems*, 42(4), 493–512.

252 *animal rights activists are not very religious*: For more information on the parallels between the animal rights movement and religion see Herzog, H. A. (1993). "The movement is my life": The psychology of animal rights activism. *Journal of Social Issues*, 49, 103–119.

253 *differences in ethical ideologies*: Galvin, S. L., & Herzog, H. A. (1992). Ethical ideology, animal rights activism, and attitudes toward the treatment of animals. *Ethics & Behavior*, 2(3), 141–149.

259 *the face of terrorism has changed over the past twenty years*: Jones, S. G., Doxsee, C., & Harrington, N. (2020, June) The escalating terrorism problem in the United States. *CSIS Briefs*. https://csis-website-prod.s3.amazonaws .com/s3fs-public/publication/200612_Jones_DomesticTerrorism_v6.pdf (Downloaded December 11, 2020).

260 *Jentsch fought back*: See Ringach, D. R., & Jentsch, J. D. (2009). We must face the threats. *Journal of Neuroscience*, 29, 11417–11418.

260 *the thinking of a dozen kinds of militant extremists*: Jasper, J. M. (1998). The emotions of protest: Affective and reactive emotions in and around social movements. *Sociological Forum*, 13(3) 397–424. Baumeister, R. F. (1999). *Evil: Inside human violence and cruelty*. New York: Henry Holt and Company, LLC. Saucier, G., Akers, L. G., Shen-Miller, S., Knezevic, G., & Stankov, L. (2009). Patterns of thinking in militant extremism. *Perspectives on Psychological Science*, 4(3), 256–271.

260 *Vlasak once told an Australian television reporter*: Best, S. (2009, May 3). Who's Afraid of Jerry Vlasak? Retrieved from http://civillibertarian.blog spot.com/2009/05/photo-and-caption-courtesy-of-guardian.html.

261 *have killed at least eleven people*: Heaveiside, J., & Mariappuram, R. (2019, May 31). The escalation of anti-abortion violence years after Dr. George Tillers murder. *Rewire Newsgroup*. https://rewirenewsgroup.com/arti cle/2019/05/31/the-escalation-of-anti-abortion-violence-ten-years-after -dr-george-tillers-murder/ (Downloaded December 7, 2020).

263 *his 1975 book* Animal Liberation: If you wanted to read one book to understand the case for animal liberation, this is it. However, for a more nuanced explanation of Singer's views see Singer, P. (1993). *Practical Ethics.* Cambridge, UK: Cambridge University Press.

264 *"The core of this book"*: Singer, P. (1975) *Animal Liberation.* New York, NY: Avon Books.

265 *legal standing for our closest relatives, apes*: See Cavalieri, P., & Singer, P. (1993). *The great ape project: Equality beyond humanity.* New York, NY: St. Martin's Griffin.

265 *Singer has acknowledged that in rare cases*: Jaschik, S. (2006, December 4). Did Peter Singer back animal research? *Inside Higher Ed* (Downloaded January 15, 2021). https://www.insidehighered.com/news/2006/12/04 /did-peter-singer-back-animal-research.

265 *the dog goes overboard*: Regan, T. (1983). *The case for animal rights.* Berkeley, CA: University of California Press (p. 324).

266 *Joan Dunayer, author of the book* Speciesism: Dunayer, J. (2004). *Speciesism.* Derwood, MD: Ryce Publishing.

266 *insects don't suffer much, he says*: Singer mentioned his views on the moral status of insects in a conversation with *New York Times* columnist Nicholas Kristof. Kristof, N. (2009, April 9). Humanity even for non-humans. *New York Times.*

266 *"Singer's disrespect for chickens"*: Dunayer, J. (2002 March-May). Letter to the editor. *Vegan Voice* (pp. 16–17).

267 *treated equally*: Singer argues that all sentient creatures deserve equal moral consideration, not equal treatment. For example, he does not believe that chimps or dogs should have the right to vote or drive.

268 *Inner Lawyer*: The idea that we all have an Inner Lawyer comes from Jonathan Haidt (2007).

268 *According to Harvard neuroscientist Joshua Greene*: Greene, J. D., Nystrom, L. E., Engell, A. D., Darley, J. M., & Cohen, J. D. (2004). The neural bases of cognitive conflict and control in moral judgment. *Neuron,* 44(2), 389–400.

271 *"predictably irrational"*: For an excellent introduction to behavioral economics see Ariely, D. (2008). *Predictably irrational: The hidden forces that shape our decisions.* New York, NY: Harper Press.

271 *dozens of types of bias*: Kahneman, D. (2011). *Thinking, fast and slow.* New York: Macmillan. Lilienfeld, S. O., Ammirati, R., & Landfield, K. (2009). Giving debiasing away: Can psychological research on correcting cognitive errors promote human welfare? *Perspectives on Psychological Science,* 4(4), 390–398.

271 *in his life boat scenario*: For a more extensive discussion of "four guys and a dog in a lifeboat" see Franklin, J. H. (2006). *Animal rights and moral philosophy.* New York, NY: Columbia University Press.

271 *OK to euthanize a permanently disabled infant*: Singer, P. (1993). *Practical*

ethics. Cambridge, UK: Cambridge University Press. For a dated but fascinating profile of Singer see Specter, M. (1999, September 6). The dangerous philosopher. *The New Yorker.*

271 *sexual interactions between humans and animals*: Singer's essay on bestiality is at Singer, P. (2001). Heavy petting. *Nerve.com.* Retrieved from http://www.nerve.com/Opinions/Singer/heavyPetting/.

274 *deep motivator for moral reflection and development*: This quote is from a email to me from Chris Diehm.

CHAPTER 10

275 *excuse for doing nothing.*: le Carré, J. (2008). *A most wanted man.* New York, NY: Scribner (p. 121).

275 *live with insecurity is the only security*: Paulos, J. A. (2002). *A mathematician plays the stock market.* New York, NY: Basic Books.

277 *The social psychologists Lucius Caviola and Valerio Capraro*: Caviola, L., & Capraro, V. (2020). Liking but devaluing animals: emotional and deliberative paths to speciesism. *Social Psychological and Personality Science*, 11(8), 1080–1088.

293 *Kwame Anthony Appiah*: Appiah, K. A. (2008). *Experiments in ethics.* Cambridge, MA: Harvard University Press (p. 199).

I spent my childhood in a steamy south Florida neighborhood with a densely overgrown vacant lot across the street that served as my private jungle. It was full of snakes, lizards, giant primordial land crabs, and a monkey who had escaped from a nearby zoo. I was always fascinated by animals, but the creatures that really grabbed me were the creepy crawlies: green anolis lizards with blood-red dewlaps, iridescent blue-tailed skinks, and the alligators that I would occasionally spot in the Miami River. My interest in animals really took off when we moved to the New Jersey suburbs. While my peers were listening to the Beatles and discussing the relative merits of hot cars, I memorized the Latin names of snakes, read scientific articles on the ecology of copperheads and chuckwallas, and dreamed of collecting trips in the backwaters of the Amazon. By the time I was fifteen, I had amassed a menagerie of a dozen or so snakes (including a small anaconda), a pair of carnivorous lizards, and a white duck named Murphy.

After high school, I attended a small college in West Virginia where I developed a taste for country music and rural living, and then moved on to the American University of Beirut, where I graduated with a degree in psychology and the desire to become an experimental psychologist. It was, however, the height of the Vietnam War. To avoid being drafted into the infantry, I enlisted as a medic and spent three years working in an army hospital in Georgia. Toward the end of my military stint, I met Mary Jean Ronan, my future wife. I went to graduate school at the University of Tennessee bent on studying cognitive psychology. But the psychology department had recently hired a young animal behaviorist named

Gordon Burghardt, who studied, of all things, the behavior of snakes. My childhood interest in reptiles was rekindled when I realized that I could get paid for doing something I really loved—watching animals. Gordon proved to be the perfect mentor. From him I learned that psychology only makes sense in the context of evolution, that interesting intellectual questions transcend academic boundaries, and to take on research projects that are fun.

Because nothing is more fun than wading knee-deep in a swamp filled with crocodilians, my first serious research involved recording the love songs of alligators. My biggest contribution to the study of animal behavior, however, was the development of a personality test for snakes. The test measured individual differences in how grouchy snakes are at birth, and it consisted of counting the number of times baby garter snakes would bite my finger during a standardized series of trials. (Because I used this technique to test many hundreds of animals, it is quite likely that I have been bitten by more snakes than any other person in human history.)

After graduate school, I was offered a teaching position at a small private college in rural western North Carolina. Mary Jean and I fell in love with the Appalachian Mountains, the people, and their culture. We raised our three children, Adam, Katie, and Betsy, in the Smokies and have never wanted to live anywhere else.

My shift from ethology to anthrozoology began shortly after we moved to the mountains, when I realized that the justifications my cock-fighting neighbors used for fighting chickens were not all that different from my rationale for eating them. I eventually became more interested in studying animal people than animal behavior, and I closed up my snake lab and began interviewing veterinary students, elephant trainers, laboratory technicians, and animal rights activists. My timing was right, and I was able to get in on the ground floor of the exciting new field of human-animal interactions.

While I was raised around cities, I am at heart a small-town guy. I have spent most of my career at Western Carolina University, where my primary job is to teach psychology to undergraduates. The balance of teaching and research offered by a regional university suits me, and the flexibility to pursue research that is decidedly outside the mainstream has made for a satisfying professional life.

I am an avid though midlevel whitewater kayaker, and I play guitar and banjo every Tuesday night with a ragtag band of mountain musicians at Guadalupe Café in Sylva, North Carolina.

When asked about the keys to happiness, Sigmund Freud is reputed to have said, "arbeiten und leben" (to work and to love). I have been fortunate in both.